ELECTRIC CONTROLS
FOR REFRIGERATION
AND AIR CONDITIONING

B. C. LANGLEY

Tarrant County Junior College
Fort Worth, Texas

Prentice-Hall, Inc.
Englewood Cliffs, New Jersey

Library of Congress Cataloging in Publication Data

Langley, B C
  Electric controls for refrigeration and air
conditioning.

    1. Refrigeration and refrigerating machinery.
—Automatic control. 2. Air conditioning—Control
3. Heating—Control. I. Title.
TP492.7.L32   621.5'6   73–6823
ISBN 0–13–247072–1
ISBN 0–13–247064–0 (pbk.)

© 1974 by
PRENTICE-HALL, INC.
Englewood Cliffs, New Jersey 07632

Current printing (last digit):
19 18 17 16

Printed in the United States of America

PRENTICE-HALL INTERNATIONAL, INC., *London*
PRENTICE-HALL OF AUSTRALIA, PTY LTD., *Sydney*
PRENTICE-HALL OF CANADA, LTD., *Toronto*
PRENTICE-HALL OF INDIA PRIVATE LIMITED, *New Delhi*
PRENTICE-HALL OF JAPAN, INC., *Tokyo*

ACKNOWLEDGMENTS

The production of this book could not have been possible without the cooperation of the various control manufacturers. In the preparation of the manuscript for this book, the following list of manufacturers and designers have been most cooperative. The author acknowledges their cooperation with much appreciation:

Arrow-Hart, Inc.
Carrier Air Conditioning Company
General Controls ITT
Honeywell, Inc.
International Register Company
McDonnell and Miller, Inc.
Motors and Armatures, Inc.
Penn Controls, Inc.
Powers Regulator Company
Ranco Incorporated
White-Rodgers Division Emerson Electric Co.

Also, to my wife, Barbara, and to Vickie, the ones who rendered their typing abilities to the manuscript, I express my appreciation.

CONTENTS

PREFACE

It is the purpose of this text to provide a simple and direct approach to the operation and application of electric controls for both refrigeration and air conditioning.

The fundamental concepts employed are applicable regardless of the changes in circuit concepts and design.

Electric Controls for Refrigeration and Air Conditioning makes use of a two-fold approach conducive to learning:

1. The theory contained in the text will assure a thorough understanding of the versatility of each control covered. The material is arranged to best provide continuity in learning. Situations are also incorporated to invoke the inquiring minds of students.
2. Practical applications of the fundamentals learned from the text may be directly employed in a student workbook and used in satisfactory laboratory situations. The workbook method has been proven to have student appeal and is a ready reference in later situations.

Even though every effort has been made to relate the principle to the actual application of controls, this is not a conclusive or exhaustive study of the constantly changing control methods employed in the field of refrigeration and air conditioning.

Upon completion of this study, the student will have the knowledge and confidence necessary to efficiently service and install control systems for refrigeration and air conditioning.

B. C. Langley

REVIEW OF MAGNETISM, ELECTRICAL CIRCUITS, AND TRANSFORMERS

Almost daily, control systems for refrigeration and air conditioning are becoming more complicated. We can no longer think of one control as an individual item, but rather, we must consider each control as a component of a complete system, having a specific reaction to a signal from another component. To understand control systems we must be familiar with the underlying principles upon which they work.

Many control devices in use today make use of magnetic fields. Permanent magnets and electromagnets are two of the types of magnets used in the operation of modern control-circuitry.

MAGNETIC FIELDS

Permanent Magnets. All types of magnets have north and south poles. All magnets have magnetic fields. When a magnetized material keeps its magnetic field for a long time it is called a permanent magnet. Hard iron or steel makes a good permanent magnet.

These magnets need no outside power source to make them work as a magnet once they have become permanently charged. Permanent magnets are used in some thermostats, switches, and valves.

Electromagnets. When a copper conductor is wound into several turns, the fields around each loop of wire combine to form a magnet when current passes through the wire. One end of the loops will be the north pole, the other one the south pole. The field strength of the coil is increased when an iron core is inserted inside the coil. Soft iron is used for electromagnet cores. The field-producing force and the permeability of the core material both have a definite relationship to the strength of the flux lines. (*Note:* Flux lines are placed in Fig. 1.1.) Soft iron has greater permeability than air.

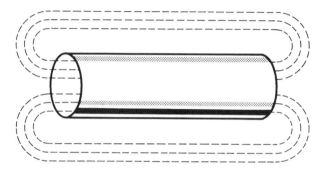

Fig. 1-1. A magnet and its field.

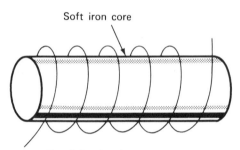

Fig. 1-2. An electromagnet.

Electromagnets are used in solenoid valves, relays, gas valves, starters, contactors, and thermocouple circuits. In these controls the magnetic field is used to convert electrical energy to mechanical energy. The magnetic field, therefore, is an important principle on which control circuits are based (Fig. 1-2).

ELECTRIC CURRENT   Electric current flow is caused by the flow of free electrons in a conductor. Electrons are very small particles within an atom. Everything is made up of atoms. Figure 1-3 shows a single atom of hydrogen.

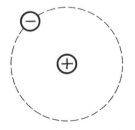

Fig. 1-3. A hydrogen atom.

The center, or nucleus, of an atom has a positive electrical charge. Electrons, which have a negative charge, revolve around the nucleus. These two oppositely charged particles are strongly attracted to each other. The nucleus of every atom has a total amount of charge equal to the number of electrons revolving around it. Different materials have different atomic structures.

The force which causes current to flow is called *electromotive force (EMF)*. EMF can be developed in several different ways. The simplest means is by chemical action on two different kinds of metals. This is known as a cell or battery. Figure 1-4 shows a typical cell or flashlight battery.

ELECTROMOTIVE FORCE

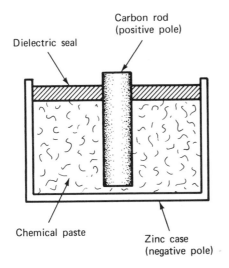

Fig. 1-4. An electrical cell.

The chemical action of the acid paste causes the materials to give up their free electrons, which will flow from the case through the circuit back to the carbon rod. The current flows in one direction—negative to positive—and produces direct current.

Because energy is removed from this cell and is not replaced, the cell will eventually be dead. These cells are small and therefore produce a small EMF. The automobile battery, or wet cell, can be recharged from an external source and will last much longer. However, it will eventually use up its source of supply.

The most popular way to develop EMF is by using the alternating current generator shown in Fig. 1-5.

In the alternating current generator a magnetic field is created by an electromagnet that is connected to an external power source. A loop of wire is rotated within the magnetic field and an EMF will thus

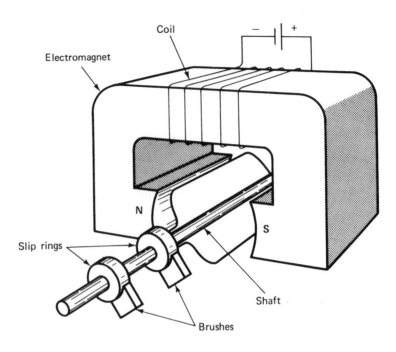

Fig. 1-5. AC generator.

develop at the ends of the coil as long as it is moving and cutting lines of force. To use this EMF we connect it to the outside of the generator by slip rings and brushes. Since the loop rotates within a magnetic field that is positive at one end and negative at the other, the EMF will be negative during one part of the cycle and positive during the other part. The current alternates between positive and negative and is called *alternating current.* Figure 1-6 shows one complete alternating cycle.

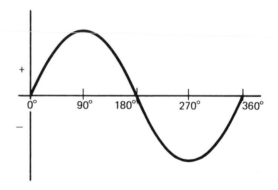

Fig 1-6. One ac cycle.

OHM'S LAW   The definite relationship between current, voltage, and resistance is known as *Ohm's Law* and may be stated as:

"The current flowing in a circuit is proportional to the voltage and inversely proportional to the resistance."

Ohm's Law can also be expressed by the equation:

$$I = \frac{E}{R}$$

which states that current (I) equals voltage (E) divided by resistance (R). Two variations of this equation are:

$$E = IR \quad \text{and} \quad R = \frac{E}{I}$$

When any two of these circuit elements are known the third may be found.

In non-mathematical language this formula is interpreted as:

as voltage is increased—current increases
as voltage is decreased—current decreases
as resistance is increased—current decreases
as resistance is decreased—current increases

SERIES CIRCUITS

If the same current flows through every part, it makes no difference how many parts or devices there are; the circuit is a series circuit.

When a circuit has a number of resistances connected in series, the total resistance of the circuit is the sum of the individual resistances. An example is shown in Fig. 1-7.

$$R_t = R_1 + R_2 + R_3 \ldots$$

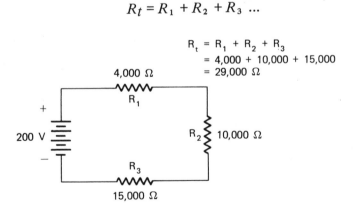

$R_t = R_1 + R_2 + R_3$
$= 4,000 + 10,000 + 15,000$
$= 29,000 \ \Omega$

Fig. 1-7. Example.

Ohm's Law applies to the whole circuit as well as to any part of it. In a series circuit the current is the same throughout the whole circuit because it follows a single path. The voltage changes, however, as each resistance in the circuit is encountered. This is known as *voltage drop* (Fig. 1-8).

The current in this circuit would be:

$$I = \frac{E}{R} = \frac{200}{29,000} = .00689 \text{ amp}$$

The voltage drop across each resistor is found by:

$$E_1 = IR_1 = .00689 \times 4,000 = 27.56 \text{ V}$$
$$E_2 = IR_2 = .00689 \times 10,000 = 68.90 \text{ V}$$
$$E_3 = IR_3 = .00689 \times 15,000 = \underline{103.35 \text{ V}}$$
$$199.81 \text{ V}$$

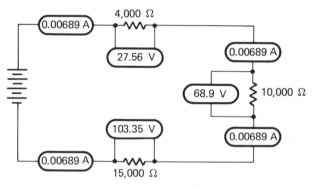

Fig. 1-8. Example.

It should be particularly noted that the sum of the voltage drops is equal to the source voltage. Our answer would have been more accurate if we had carried the current equation to more decimal places.

Kirchoff's Laws summarize the facts concerning series circuits as:

1. The sum of the voltage drops around a series circuit will equal the source voltage.
2. The current is the same when measured at any point in the series circuit.

PARALLEL CIRCUITS

When more than one component is connected to a single voltage source and they are connected side by side, they are said to be in parallel.

One must realize that when components are connected in this manner, the total resistance is decreased every time another component is added to the circuit.

A parallel circuit can be compared to a system of pipes carrying water. Two pipes will carry more water than will one. Likewise, three pipes will carry more water than two. As more pipes are added the total resistance to the flow of water is reduced, or decreased. This is illustrated in Fig. 1-9. In a circuit with resistances in parallel the total resistance is found by:

$$R_t = \frac{1}{\dfrac{1}{R_1} + \dfrac{1}{R_2} + \dfrac{1}{R_3} + \ldots}$$

or if only two resistors are in parallel (see Fig. 1-10).

$$R_t = \frac{R_1 \times R_2}{R_1 + R_2}$$

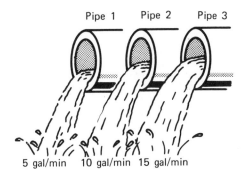

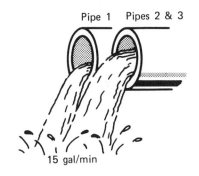

Fig. 1-9. Example.

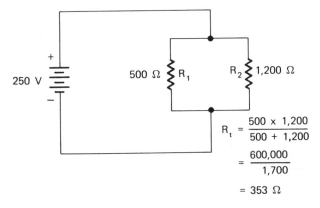

$$R_t = \frac{500 \times 1,200}{500 + 1,200}$$
$$= \frac{600,000}{1,700}$$
$$= 353 \ \Omega$$

Fig. 1-10. Example.

With resistances in parallel, the total resistance is always less than the value of the least resistor present. This is because the total current is always greater than the current in any individual resistor. Therefore, in the following example the total resistance of the circuit must be one-third of the resistance of a single resistor. If we assign values of 30 ohms to each resistor in Fig. 1-11 then:

$$R_t = \frac{R \text{ (value of one resistor)}}{N \text{ (number of resistors in network)}} = \frac{30}{3} = 10 \text{ ohms}$$

The applied voltage is the same for each branch in a parallel circuit, because all the branches are connected across the same voltage source, as in Fig. 1-11.

Therefore, using Ohm's Law the currents are:

$$I = \frac{E}{R} = \frac{6V}{30\Omega} = .2 \text{ amp across } R_1$$
$$= \frac{6V}{30\Omega} = .2 \text{ amp across } R_2$$
$$= \frac{6V}{30\Omega} = .2 \text{ amp across } R_3$$

The total current flow through the network would be the sum of the individual branch currents or:

$$I_t = I_{R_1} + I_{R_2} + I_{R_3}$$
$$= .2 + .2 + .2 = .6 \text{ amp}$$

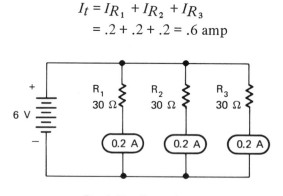

Fig. 1-11. Example.

To summarize, we give these laws concerning parallel circuits:

1. The voltage across all branches of a parallel network is the same.
2. The total current is equal to the sum of the individual branch currents.

A comparison should be made between these laws and the ones which apply to series circuits.

Series-parallel circuits are a combination of both series and parallel circuits. They can be fairly simple and have only a few components, but they can also have many components and be quite complicated.

In any circuit, there are certain basic factors in which you should be interested. From what you have learned about series circuits and parallel circuits, you know that these factors are (1) the total current from the power source and the current in each part of the circuit, (2) the source voltage and the voltage drops across each part of the circuit, and (3) the total resistance and the resistance of each part of the circuit. Once these factors are known the others can be easily calculated.

When calculating either type of circuit, series or parallel, you will use only the rules that apply to that type. On the other hand, in a series-parallel circuit some components are connected in series and some are connected in parallel. Therefore, in some sections of a series-parallel circuit you have to use rules for series circuits, and in other sections the rules for parallel circuits apply.

You can see then, that before you can analyze or solve a problem involving a series-parallel circuit you must be able to recognize which parts of the circuit are series-connected and which parts are parallel-connected. If the circuit is simple this is sometimes obvious. However, there are times when the circuit will have to be redrawn, setting it into a form that is easier to recognize, such as in Fig. 1-12.

SERIES-PARALLEL

CIRCUITS

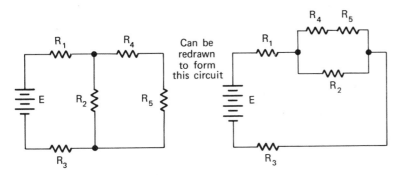

Fig. 1-12. Example.

This is the first of four steps involved in computing the resistance of a series-parallel circuit.

Step 2 is to combine the series branch circuits, for example: (Fig. 1-13)

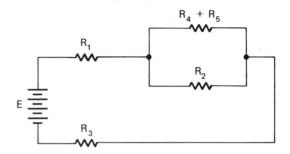

Fig. 1-13. Example.

Step 3 is to find the total resistance of the parallel circuit containing R_4, R_5, and R_2 (Fig. 1-14).

$$R_t = \frac{1}{\frac{1}{R_4} + \frac{1}{R_5} + \frac{1}{R_2}}$$

The circuit now appears as:

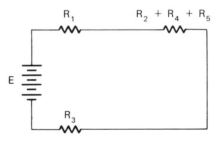

Fig. 1-14. Example.

Step 4 is to compute the series circuit by Ohm's Law (Fig. 1-15).

$$R_t = R_1 + R_2 + R_3 \text{ ...}$$

The circuit now appears as:

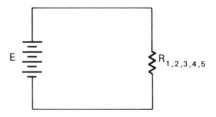

Fig. 1-15. Example.

Remember that you usually cannot calculate all of the currents or all of the voltages in a series-parallel circuit by using only the total current and applied voltage. You have to work around the circuit load by load and branch by branch, finding the current through and the voltage across each load or branch before moving on to the next. As you acquire more experience and practice you will develop your own short cuts, which will enable you to eliminate some of the work involved in calculating series-parallel circuits.

POWER AND ENERGY

Power is defined as the rate of doing work. It is equal to the voltage multiplied by the current. The unit of power is the watt. The formula for figuring power is:

$$P = E \times I$$

By substituting Ohm's Law equivalents for E and I we get these additional formulas:

$$P = \frac{E^2}{R} \quad \text{and} \quad P = I^2R$$

Energy is the work which electricity does and is measured in *watthours*. The *watthour* formula is:

$$W = PT$$

Where: W = energy in watts
   P = power in watts
   T = time in hours

INDUCTANCE

As we learned earlier, a magnetic field contains lines of force. When a current is passed through a straight piece of wire there is a small concentration of lines of force, as shown in Fig. 1-16.

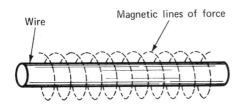

Wire

Magnetic lines of force

Fig. 1-16. Lines of force.

If we form this wire into a coil we will get a stronger concentration of lines of force (Fig. 1-17).

The magnetic field which is created by a current flowing through the coil induces a current in the coil itself as the magnetic lines of force

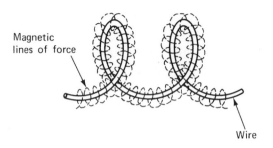

Magnetic
lines of force

Wire

Fig. 1-17. Magnetic field in a coil.

cut across the conductor. This induced current will be opposite in polarity to the applied current. When the applied current is increased the induced current is also increased. This action tends to oppose a change in current. As the applied current starts to build up it is opposed by the induced current. So, in an inductive circuit, the change in current always lags behind the change in voltage.

ALTERNATING CURRENTS

To understand alternating currents we must first understand the ac phase. *Phase* is the time interval between the instant one thing occurs and the instant when a second and related thing takes place. To do this we divide alternating current into cycles. Each ac cycle takes the same amount of time as all other cycles of the same frequency. Alternating current is divided into 360° as shown in Fig. 1-18.

The ac wave form is called a *sine wave*. As you will note, the current is positive through one half the cycle and negative through the other half.

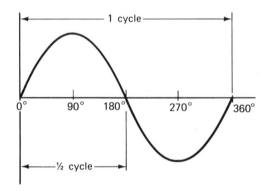

Fig. 1-18. AC cycle.

TRANSFORMERS

A *transformer* may be defined as a device used to transfer electrical energy from one circuit to another.

Basically, a transformer consists of two or more coils wound around one laminated core, so that the coupling between them approaches unity (so that all the lines of magnetic flux of one coil will cut across all the turns of a second coil). These devices have no moving parts and require very little maintenance. They are simple, rugged, and efficient. Transformer diagrams are shown in Fig. 1-19.

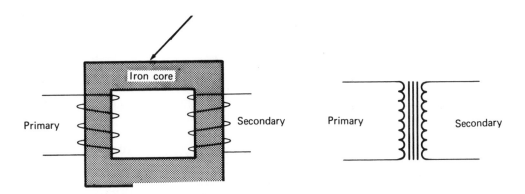

Fig. 1-19. Symbolic sketch of transformer and schematic symbol.

In operation, the primary circuit draws power from the source, and the secondary circuit delivers the power to the load. The power transferred from the primary winding to the secondary winding is determined by the current flowing on the secondary circuit. The current flow in the secondary circuit depends on the power required by the load. If the load has low resistance and requires a great deal of power, high current will flow in the secondary circuit. This high current causes a decrease in the EMF of the magnetic field that is necessary for the high current flow in the secondary circuit. Thus, the transformer regulates the transfer of power from the source to the load in response to the load requirements.

The transformer is used in power transmission to convert power at some value of current and voltage into the same power at some other value of current and voltage. This can be done because, with a given voltage in the primary circuit, the secondary voltage depends on the number of turns in the secondary winding as compared to the number of turns in the primary winding.

Careful consideration should be given to the selection of a transformer that is to power a low voltage control system.

For example, inductive devices such as contactors, relays, solenoid valves, and motors require more power on starting than during steady operation. A transformer delivers the maximum possible inrush current to a load when the transformer impedence equals the impedence of the load. The performance of an inexpensive transformer that is properly matched to the load can be equivalent to that of a more costly, unmatched transformer.

QUIZ 1

1. Name two types of magnets used in modern control devices.

2. All magnets have a _____ pole and a _____ pole.

3. A permanent magnet is one that keeps its _____ for a long time.

4. What type of material makes a good permanent magnet?

5. Name four uses for permanent magnets in modern controls.

6. In an electromagnet the field strength is increased when _____. _____ .

7. The permeability of soft iron is _____ than air.

8. How is an electromagnet made?

9. The magnetic field is increased when the current through the coil is _____ .

10. A magnetic field can be used to convert _____ energy into _____ energy.

11. Name four uses for electromagnets in modern controls.

12. Give the equation for Ohm's Law.

13. In a series circuit, the total resistance is _____ .

14. The formula for calculating current is _____ .

15. State the four laws concerning series and parallel circuits.

16. The total resistance in a parallel circuit _____ with the addition of resistances.

17. In a parallel circuit, the applied voltage is the same for each _____ .

18. In a series circuit, the sum of the voltage drops equals the _____ .

19. Define a transformer.

20. A transformer is used in power transmission to _____ .

2

MAGNETIC STARTERS
AND CONTACTORS

In the refrigeration and air conditioning industry, the compressor motor represents the largest switching load for the control system. The various fan motors, water pumps, and other machinery wired in parallel with the compressor motor also add to the load requirements.

There are several methods of controlling the various loads but we will discuss only the starter and contactor at this time.

DEFINITIONS

The National Electrical Manufacturers Association defines starters and contactors as follows:

A *starter* is an electric controller for accelerating a motor from rest to normal speed.

A *contactor* is a device for repeatedly establishing and interrupting an electric power circuit.

Contactors have many common features. Each has electromagnetically operated contacts which make or break a circuit in which the current exceeds the operating circuit current of the device. Each may be used to make or break voltages which differ from the controlling voltage. A single device may be used to switch more than one circuit. Thus, starters and contactors offer the following features as elements of an electric circuit:

1. Isolation between the control circuit voltage and the controlled voltage made complete.

2. High power gain of controlled to controlling power.

3. Ability to control a number of operations from one device with a single control voltage. Also, complete electrical isolation can be obtained between controlled voltages.

A contactor is used for switching heavy current, high voltage, or both (Fig. 2-1).

A motor starter may consist of a contactor employed as a means of switching power to a motor. A starter, however, usually has additional components, such as overload relays and holding contacts. These components may also include step resistors, disconnects, reactors, or other hardware required to make a more sophisticated starter package for large motors (Fig. 2-2).

Fig. 2-1. Three-pole contactor. (Courtesy of Arrow-Hart, Inc.)

Fig. 2-2. Five-pole starter. (Courtesy of Arrow-Hart, Inc.)

OPERATION   Different manufacturers build starters and contactors with a variety of armatures. Some connect armatures to the relay frame with hinges or pivots, and others use slides to guide the armature. Except for this difference their basic operation is the same; therefore, we will discuss only the hinge type (Fig. 2-3).

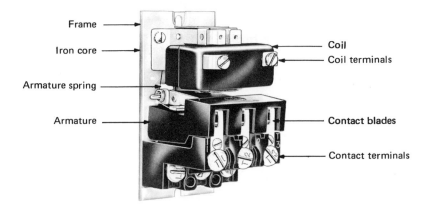

Fig. 2-3. Contactor components. (Courtesy of Arrow-Hart, Inc.)

The armature is the moving part of the starter or contactor. It is hinged on one end to the frame and has a movable contact assembly on the other end. A spring keeps the armature pulled away from the electromagnet when the coil is de-energized. The metallic armature is easily magnetized by the lines of flux from the energized electromagnet. The magnet consists of a coil wound around a laminated iron core; this becomes an electromagnet and pulls the armature toward it when the coil is energized.

COIL

Coil characteristics depend on the wire and the method of winding. The potential-wound type responds to some value of voltage to pick up the armature. The current-wound type responds to some value of current. Coil terminals connect the coil to the control voltage wiring through a switching device such as a thermostat or pressure control.

CONTACTS

The type of load is very important when sizing contactors and starters for particular application. To properly match the starter or contactor to the application, the engineer must consider the different characteristics of inductive and noninductive (resistive) loads.

In an ac circuit with an inductive load such as an electric motor, the initial inrush current is always much higher than the normal operating current. Before the motor obtains sufficient speed the back EMF is very low, allowing maximum current to flow. As the motor gains speed, the back EMF increases and opposes the current flow, reducing it to the normal running current. Starter and contactor contacts, therefore, must not only withstand the normal running current, but also the starting inrush of current. A good table of contact ratings lists both full-load and locked-motor current ratings. See Table 2-1.

Table 2-1

Single-Phase Motor		Three-Phase Motor	
Compressor Size (Tons)	Motor Current (Amperes)	Compressor Size (Tons)	Motor Current (Amperes)
2	18	3	18
3	25-30	4	25-30
4	30-40	5	30-40
5	40-50	7½	40-50

Contacts are made of silver cadmium alloy, which gives the highest resistance to sticking. The contacts are bonded to a strong backing member. They operate coolly even at loads of 25 per cent above their ratings.

Pole Configuration. Starters and contactors normally use two- or three-pole sets (Fig. 2-4) to engage the compressor motor, depending on the current required for the load (single phase or three phase). These devices may be purchased with only the necessary poles or with extra poles as well. These extra contacts are either unused or used as auxiliary contacts.

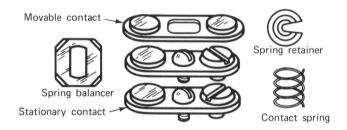

Fig. 2-4. Set of contacts constituting one pole.

Auxiliary contacts are normally used to complete an interlock circuit such as another contactor connected to a water pump, fan motor, or similar device, or to complete a circuit while in the de-energized position. These contacts are usually only for pilot duty and are not intended to withstand the heavy current used by the main load.

These devices are usually mounted on the side of a contactor, thereby completing the definition requirements of a starter. Overload relays may also be conveniently mounted elsewhere in the circuit. They are safety devices used to protect the main load from damage caused by excessive current.

Operation: Overload relays operate on the principle that current produces heat. The main load is directed through the relay by use of a resistance wire. If the current exceeds the rating for the resistance wire, additional heat is radiated to a bimetal switch. When sufficient heat is generated the bimetal switch opens a set of contacts in the control circuit and allows the starting device to de-energize, thereby stopping the motor. The bimetal switch must be manually reset before the starter will again complete the electrical circuit to the load (Fig. 2-5).

OVERLOAD RELAYS

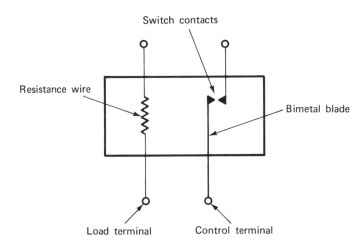

Fig. 2-5. Representative overload relay.

Starters and contactors are used for many purposes. Today, however, over 90 per cent of those manufactured are used by the refrigeration and air conditioning industry. They are used for the starting and operation of refrigeration compressors, large fan motors, and water pumps. Starters and contactors prove to be the most useful when heavy loads are to be started and when the control voltage is different from the power voltage. When we include electric heat and flame safeguard applications along with all the other uses in the refrigeration and air conditioning industry, we have completely covered the uses of contactors and starters (see Fig. 2-6).

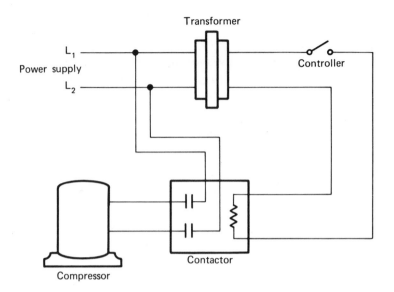

Fig. 2-6. Typical diagram using a contactor.

QUIZ 2

1. How are starters and contactors operated?
2. How many circuits can a starter or contactor switch?
3. How can a contactor be differentiated from a starter?
4. What causes an overload relay to function?
5. How are auxiliary contacts used?
6. What are contacts made of?
7. Where are the majority of starters and contactors used?
8. How many poles are required for three-phase motors?
9. A pole consists of two _____ contacts and two _____ contacts.
10. The two methods of winding contactor and starter coils are _____ and _____ .
11. Two ways of connecting an armature to the base are _____ and _____.
12. When a starter or contactor is used, the control circuit is _____ from the controlled circuit.
13. A starter or contactor is used to _____ .
14. A contactor armature is easily _____ .
15. Name three common characteristics of contactors.

3

MAGNETIC RELAYS AND THERMAL RELAYS

The relay enjoys the greatest demand in the controls industry. It has many uses in refrigeration and air conditioning control systems. Each year new and imaginative requirements are specified by equipment manufacturers.

A *relay* is a switching device which operates from an electrical input signal to affect the operation of devices in the same circuit or other circuits.

Relays, as studied here, are devices designed especially for the automatic control of one or two speed fan motors in heating, refrigeration, and air conditioning, as well as for heating and cooling control. The application of relays will end only with an end to the imagination. It would be almost impossible to cover all applications for relays at one time; therefore, we will discuss only a small representation (Fig. 3-1).

DEFINITION

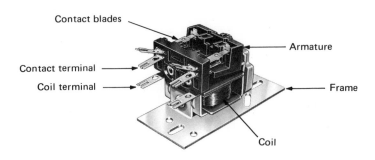

Fig. 3-1. Magnetic relay. (Courtesy of Honeywell, Inc.)

OPERATION   The operation of magnetic relays is almost the same as the operation of starters and contactors. If their operation is not clear a review is suggested at this time. The main difference in operation is the temperature of the relay coil, especially when the relay is enclosed.

A coil is limited by the amount of heat it can dissipate in a given time to the ambient air or to the core of the magnet. Increasing the operating voltage causes an increase in temperature.

The outer windings of a coil operate more coolly than the inner windings because they are closer to the surrounding air and give up more heat. This temperature difference causes hot areas to develop inside the coil; this deteriorates the coil insulation, causing shorted windings.

POLE
CONFIGURATION   Relays may be obtained with almost any type of pole arrangement imaginable. The main arrangements are: normally open (NO), normally closed (NC), single pole-double throw (SPDT), double pole-single throw (DPST), double pole-double throw (DPDT), and variations involving the above-mentioned arrangements. For symbols of these arrangements see Table 3-1.

Table 3-1. Relay symbols for electrical diagrams.

Pole Form	Symbol
SPST, N.O.	
SPST, N.C.	
SPDT	
DPST, N.O.	
DPST, N.C.	
DPST, N.O. and N.C.	

Some relays have normally closed contacts and open a circuit when the relay operates. Circuit diagrams always show relays in the de-energized position.

Electromagnetic Relays. When the control circuit is complete and power is delivered to the coil, the electromagnet pulls the armature toward it. As the normally open contacts are closed (made) the electrical circuit to the other devices is completed. When the control power to the relay coil is interrupted, the electromagnet loses its power, permitting the armature spring to pull the armature away from the coil, and thus breaking the connection to the output circuit devices.

Fan Relays. The fan relay is one of the most used electromagnetic relays in the refrigeration and air conditioning industry. Its main purpose is to bypass the winter fan control and operate the fan motor for air conditioning or ventilation. This relay may be purchased alone or with a transformer attached. The contacts of this relay should be heavy enough to withstand the current used by the fan motor. It has normally open contacts (Fig. 3-2).

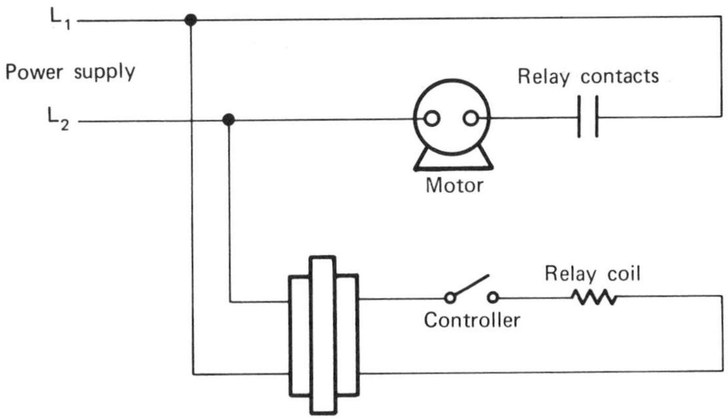

RELAYS IN THE CONTROL CIRCUIT

Fig. 3-2. Typical schematic relay circuit.

Lock-out Relay: The lock-out relay is used as a compressor motor safety device. The relay coil voltage is the same as the contactor coil voltage, and the contacts may be of the pilot duty type. It is wired into the circuit with its normally closed contacts in series with the starter or contactor coil. The relay coil is wired parallel to the starter or contactor coil (Fig. 3-3). If any of the other safety devices, such as pressure switches, overloads, or the like, stop the compressor, the lock-out relay keeps it from restarting until it is reset by turning off the electrical power to it.

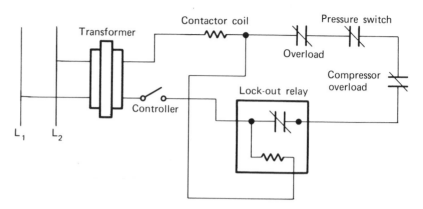

Fig. 3-3. Typical lock-out relay diagram.

Thermal Relays.   The thermal relay contacts and terminals are the same as in the electromagnetic relay. The method of pulling the movable contacts against the fixed contacts is the major difference between the electromagnetic and the thermal relays. The thermal relay utilizes a bimetal blade with a heater coil wound around it. The heater leads connect to the control power supply through the low voltage controller. When the controller demands, the heater warms the bimetallic element, causing it to bend toward the stationary contact to complete the output circuit. When the controller opens the control circuit to the heating element, the bimetal element cools and moves away from the stationary contact, breaking the circuit to the external device. (Fig. 3-4).

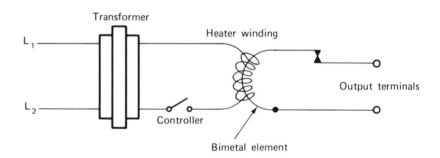

Fig. 3-4. Thermal delay relay circuit.

Because there is an inherent delay in making the contact, these relays are often used as timing devices or time delay relays.

CONTACTS   Most relay contacts are silver cadmium oxide mounted on beryllium copper blades for longer life and low resistance. Some manufacturers build special models available with gold flash contacts for powerpile systems.

These relays are operated by either current or voltage (potential). Current relays can be either coil or thermal element actuated, but voltage relays depend solely upon electromagnetic coils.

Voltage Relay. This relay may be recognized by its resistance coil wound with very fine wire. The coil of the voltage relay is connected in parallel with the starting winding of the compressor.

Voltage relays are normally used on high starting-torque motors, but they may also be used on low starting-torque motors. The relay contacts are normally closed.

This is normally a three terminal relay, but it may also have auxiliary (binding) terminals. The terminals numbered 1, 2, and 5 are the operating terminals while 3, 4, and 6 are binding terminals, if these are included in the relay (Fig. 3-5).

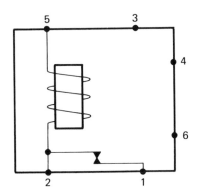

Fig. 3-5. Symbolic voltage relay.

When wiring the relay into the system, terminal 1 and the run winding terminal on the motor are connected to the same supply line. The starting capacitor is installed in this line (Fig. 3-6). Terminal 5 and the motor common winding terminal are connected to the same supply line. Terminal 2 is connected to the motor starting winding terminal.

Operation. When power is first supplied to the motor, the contacts of the starting relay are closed. In the closed position there is an electrical out-of-phase condition between the motor windings. During the initial start up there is no back-EMF and maximum current is flowing. As the motor speed increases a back-EMF is produced causing an increase in voltage. When the voltage is increased to the pick-up voltage of the motor starting relay, the armature will actuate opening the contact points. At this point the out-of-phase condition is relieved, allowing the motor to function under normal operating conditions.

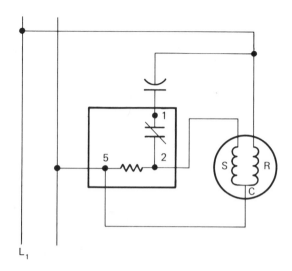

Fig. 3-6. Typical voltage relay wiring diagram.

Current Relay.   The coil (or thermal element) of this type of relay is connected electrically in series with the running winding of the motor. The coil of the current relay can be recognized by the low resistance, heavy wire with which it is wound.

Current relays are used mostly on low starting-torque motors. The contacts of this relay are normally open.

Both three and four terminal relays are available. Some of them have binding terminals. Three terminal current relays have switch connections from *L* to *S* and the coil between *L* and *M* (Fig. 3-7). Four terminal relays have switch connections from 3 to *S* and the coil between 2 and 4 and M (Fig. 3-8). Four terminal relays are often used with three terminal overloads and provide a convenient method of connecting starting capacitors.

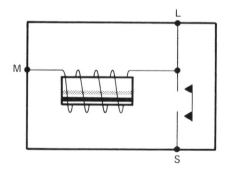

Fig. 3-7. Symbolic three-terminal current relay.

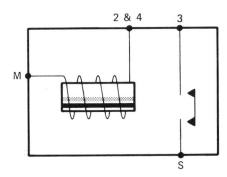

Fig. 3-8. Symbolic four-terminal current relay.

Operation. When power to the motor is first turned on, the contacts of the coil-type current relay are open. The high inrush or current through the electromagnetic coil closes the contacts, thereby producing the out-of-phase in the motor needed to start it. As the back-EMF increases, the current decreases, allowing the relay contacts to open and remove the out-of-phase condition. The motor then operates as usual.

The thermal element (hotwire) type of current relay (Fig. 3-9) operates much the same as the voltage relay. The only exception is that the main current is routed through the relay by use of a resistance wire. The heat generated during the maximum current inrush actuates a bimetal warp switch. The switch opens the contacts and removes the out-of-phase condition in the motor. Also, the main contacts remain closed and open only in an overcurrent condition to stop the compressor.

A review of the counter-EMF in electric motors may be needed before this is completely understood.

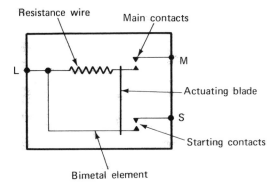

Fig. 3-9. Symbolic thermal-type current relay.

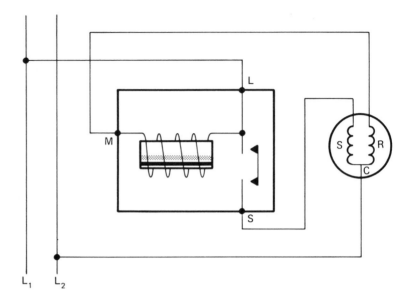

Fig. 3-10. Typical current relay wiring diagram.

QUIZ 3

1. The relay operates much like _____ and _____.

2. What is the difference between contactors and relays?

3. Are relays used very much in the control of refrigeration and air conditioning?

4. Draw the symbol for a DPST relay.

5. What operates the switch in a thermal relay?

6. What material are relay contacts made of?

7. Name two types of motor starting relays.

8. Give another name for a thermal element motor starting relay.

9. Which current relay operates very much like the voltage starting relay?

10. What is the purpose of a motor starting relay?

4

SOLENOID VALVES, REVERSING VALVES, AND COMPRESSOR UNLOADERS

These controls are used in refrigeration and air conditioning to direct, channel, or directly affect the flow of fluids.

The automatic control of refrigerants, brine, gas, or water depends frequently on the use of one or more of these valves. They may be used individually or together in a system to complement each other and provide an ultimate in control systems.

A *Solenoid valve* is a device made by putting two separate devices into one control. A solenoid is a coil of wire (Fig. 4-1) which when carrying electric current has the characteristics of an electromagnet. When this coil is attached to the stem of a valve a plunger is pulled into the coil providing the desired action of the valve. Hence it is called a solenoid valve.

DEFINITIONS

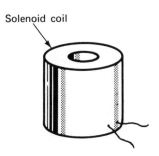

Fig. 4-1. Solenoid coil.

A *Reversing valve* (four way valve) provides four flow paths—two at a time. Two separate flows can be diverted in two directions by virtue of the valve operation.

A *Compressor unloader* loads or unloads compressor cylinders by allowing discharge gas to bypass to the suction side of the cylinder through a bypass port, thus controlling the capacity of the compressor.

Solenoid valves are probably the most used device of all controls in the refrigeration and air conditioning industry. Because they are electrically operated, and perform many functions of a manually operated valve, they are very desirable for automatic operations.

These valves can be made to perform many functions merely by the turning on or off of an electrical circuit, which is usually accomplished by using a thermostat. When the electrical circuit to the solenoid coil is completed an electromagnetic field (Fig. 4-2) is created which pulls the plunger into the center of the coil. By attaching this plunger to a valve stem the valve can be made to open or close upon energizing or de-energizing the electrical circuit.

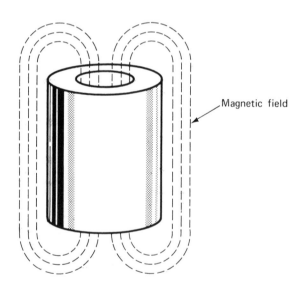

Magnetic field

Fig. 4-2. Magnetic field.

There are two general types of solenoid valves: (1) the direct acting (Fig. 4-3) and (2) the pilot operated (Fig. 4-4).

In the direct acting type (Fig. 4-3), when the circuit is completed to the coil *A* the energized coil pulls the plunger *B* upward, lifting the valve disc *C* from the valve seat *D*.

This allows the fluid to pass through the valve until such time as the circuit is broken to the coil *A*. The plunger *B* will then drop, and the valve *C* will rest on the seat *D*, shutting off the flow of gas. The valve spring *E* assures positive closure.

Valves that are equipped with manual operation have a knob on

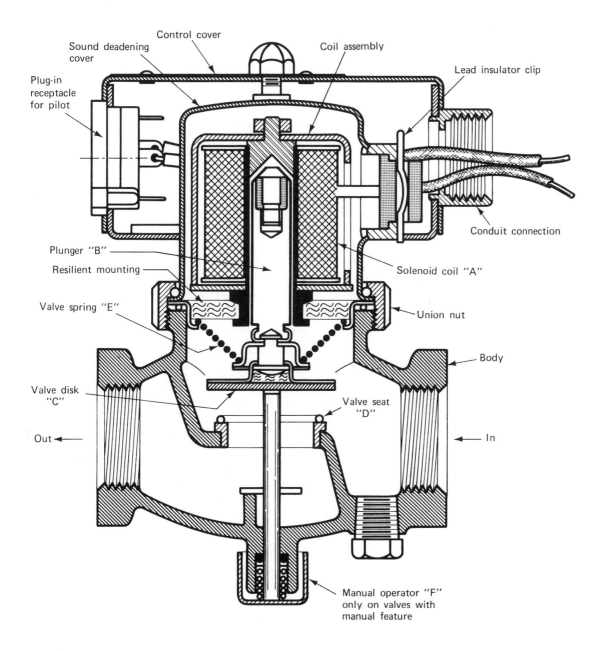

Plug-in
receptacle
for pilot

Sound deadening
cover

Control cover

Coil assembly

Lead insulator clip

Conduit connection

Plunger "B"

Resilient mounting

Valve spring "E"

Solenoid coil "A"

Union nut

Body

Valve disk
"C"

Valve seat
"D"

Out

In

Manual operator "F"
only on valves with
manual feature

Fig. 4-3. Direct acting solenoid valve.

the bottom of the valve body. In case of power failure the valve may be
opened by sliding the instruction sleeve off the knob and pushing the
knob upward with the fingers. Then the knob is rotated one quarter-
turn in either direction to hold the valve *C* in the open position.

To manually close the valve, rotate the knob one quarter-turn and
the manual operator will return to its normal position. The force of the
valve spring *E* will return the valve to a fully closed position. *Note:*
when on manual operation, the control does not return to the automatic operation when the power is restored.

Pilot operated solenoids (Fig. 4-4) are usually made in the larger sizes. The plunger does not directly open the main valve seat in this type of valve. When the coil *A* is energized the plunger *B*, which has two seats, moves from seat *C* to seat *D*. The pilot port *E* is thus closed and the bleed port *F* is opened allowing the fluid to escape to the outlet side of the valve. The pressure now is the same on both sides of the diaphragm *G* allowing the spring *H* to open the main seat and allow the fluid to flow straight through the valve.

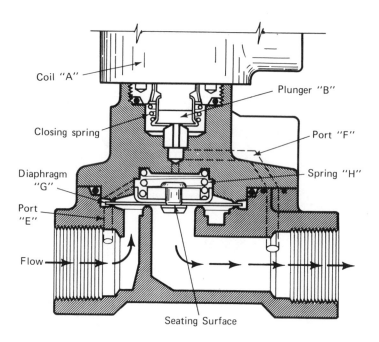

Fig. 4-4. Pilot operated solenoid valve.

All solenoid valves are made on these basic principles. There are, however, a few exceptions in the mechanical construction of these valves. Some examples are the lever valve, the general purpose two- or three-way valve, and the four-way valve. The proper valve should be chosen for the specific installation.

For the solenoid valve to function properly, the following general rules should be kept in mind when the valve is installed:

1) Arrows on the valve body indicate the proper direction of flow through the valve.
2) It is advisable to use new pipe, properly chamfered and reamed, when making connections. Be careful when using pipe dope.
3) The ambient air temperature immediately surrounding the solenoid valve must not exceed 125° F.

Reversing valves are arranged for various nominal tonnage capacities for the automatic operation of heat pump air conditioning systems, using temperature controls, in addition to the valves. Size should be according to the manufacturer's specification. These valves are hermetically constructed and are pressure-differential operated. Their operation is controlled by an energized or de-energized solenoid coil secured over a three-way pilot valve with a lock nut, integral with the main valve.

The valves are instantaneous in reversing against running pressures and operate wholly on pressure differential between the high and the low sides of the refrigeration system under full pressure within its listed capacities.

The refrigerant gas path is schematically diagrammed through the main valve showing the sliding port at a position over two tube openings as it transfers both refrigerant coils between the operating phases of cooling, de-icing and heating.

The solenoid coil is not energized in the normal cooling cycle (Fig. 4-5), and the refrigerant flows through its conventional cycle.

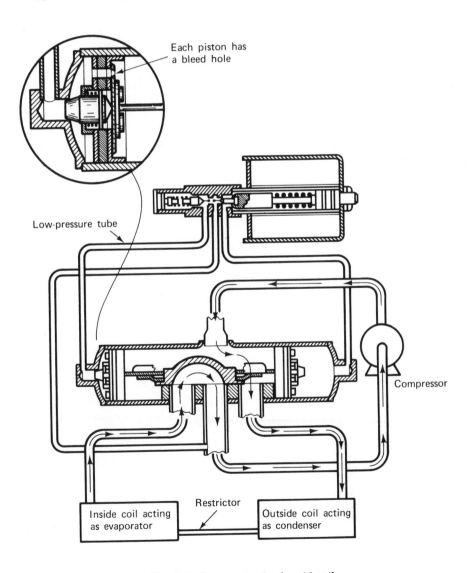

Fig. 4-5. De-energized solenoid coil.

In the heating cycle, (Fig. 4-6) the solenoid coil is first energized to operate the pilot valve plunger. This closes the left port with one needle valve and keeps the right port open with the other (Fig. 4-6). The pressure differential, created between the two main valves and the chambers by pilot valve action, instantly causes two pistons to move the sliding post. Both end chambers soon arrive at equal pressures within an operating phase. However, this condition is instantly changed by action of the pilot valve in response to the temperature control action.

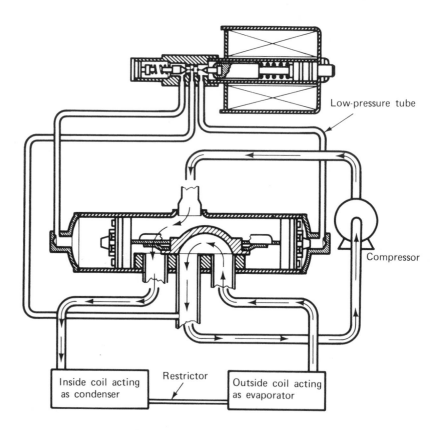

Fig. 4-6. Energized solenoid coil.

Compressor unloaders are available on all large compressors. They are usually mounted on the cylinder head of one or more cylinder banks. They are self-actuated, suction pressure or electric solenoid controlled, and discharge pressure operated. Valve operation is such that controlled cylinders do not load up until a differential of approximately 25 psi is established between suction and discharge pressure.

Each control valve loads or unloads one compressor cylinder bank by allowing discharge gas to pass to the suction side of the cylinder through a bypass port. Unloaded cylinders operate with no pressure differential, and thus consume little power.

The cylinder load point, on suction pressure controlled valves, is adjustable from approximately 0 psig to approximately 85 psig. The pressure differential between the cylinder load-up point and the cylinder unload point is adjustable from about 5 psig to 20 psig.

Electric solenoid valves (Fig. 4-7) unload controlled cylinders in response to external thermostat or pressurestat. Pressure differential for complete unloading range varies with the control device used.

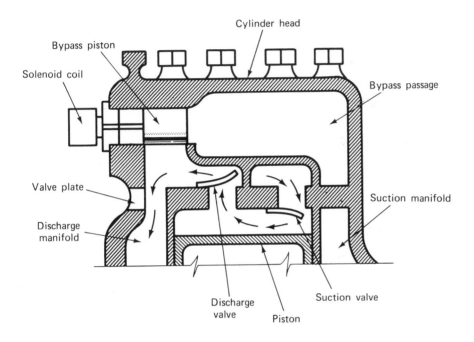

Fig. 4-7. *Loaded cylinder as used on carrier compressor.*

Energizing the solenoid valve (Fig. 4-7) unloads one cylinder bank, and de-energizing the solenoid valve loads the cylinder bank.

Loaded Operation. When the suction pressure is above the control point the poppet valve will close. The discharge gas bleeds into the valve chamber, and the pressure closes the bypass piston and the cylinder bank loads up. Discharge gas pressure forces the check valve open permitting gas to enter the discharge manifold (Fig. 4-7).

Unloaded Operation. When the suction pressure falls below the valve control point, the poppet valve will open. The discharge gas now bleeds from behind the bypass piston to the suction manifold. The bypass piston opens and the discharge gas is recirculated back into the suction manifold and the cylinder bank is unloaded. A reduction in the discharge pressure will cause the check valve to close, isolating the cylinder bank from the discharge manifold (Fig. 4-8).

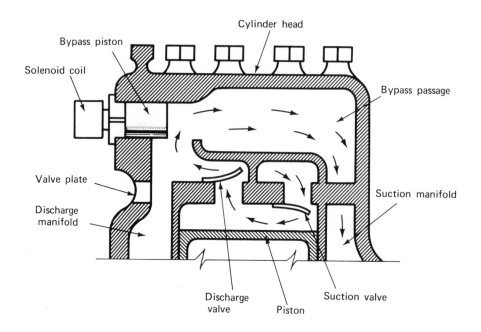

Fig. 4-8. Unloaded cylinder as used on carrier compressor.

COILS   Most solenoid coils are molded in an epoxy resin. The coil assembly is not waterproof, but it is moisture resistant and will not usually break down electrically when high voltage is applied between the coil and the valve body after prolonged immersion in water at room temperature. The wire winding in these coils is like most other electromagnetic coils.

Solenoid valves may be used as a main liquid line control valve on multiple systems as insurance against liquid flood back on compressor start up. They may also be used to prevent liquid from entering the low side by leaking through the expansion valve during the off cycle (Fig. 4-9). Often they are used to control the flow of refrigerant to individual evaporators, all of which are connected to the same compressor. These are only samples of uses of an indispensible control in the refrigeration industry.

Reversing valves are used in two main areas: (1) in heat pump applications, and (2) in hot gas defrost systems in refrigeration work.

Compressor unloaders are used as they are named. They are used mostly in large systems to prevent the evaporator from frosting and to reduce power consumption during low load conditions.

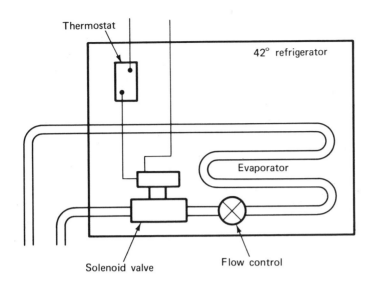

Fig. 4-9. Refrigerant flow controlled by a solenoid valve.

QUIZ 4

1. A solenoid uses the _____ principle for operation.
2. Compressor unloaders are used to _____ power consumption.
3. Evaporator frosting may be prevented by the use of _____.
4. A _____ diverts the _____ flow.
5. Solenoid valves perform like _____ .
6. Two types of solenoid valves are _____ and _____:
7. Reversing valves are _____ against running pressures.
8. When a compressor unloads the suction pressure _____.
9. The compressor cylinder load point is adjustable.
10. A compressor cylinder bank unloads when the solenoid is _____ .

5

PRESSURE AND
OIL FAILURE CONTROLS

Whenever an electric motor is stalled or overloaded the motor draws current many times its full load rating. If the condition is allowed to continue the motor windings overheat. The least that will happen is that the insulation on the motor windings will be destroyed, rendering the motor inoperative.

A motor or compressor may be protected from overcurrent damage, or bearing damage, by use of pressure and oil failure controls.

Pressure controls are switching devices used to stop the compressor motor when the pressures reach a predetermined point.

DEFINITIONS

Oil failure controls are switching devices that give dependable protection against major breakdowns on pressure lubricated refrigeration compressors by guarding against low lube oil pressure.

In general, a low pressure control is connected to the low side of the compressor and is adjusted to stop the compressor if the low side pressure drops below a desired level.

OPERATION

The main purpose of the low pressure control in air conditioning is to prevent the temperature of the evaporator coil from falling below the temperature at which frost would occur. It may have either manual or automatic reset (Fig. 5-1).

It is used often in refrigeration work as a temperature control, and sometimes as a defrost control, depending on the cut-in and cut-out settings. (See Table 5-1.)

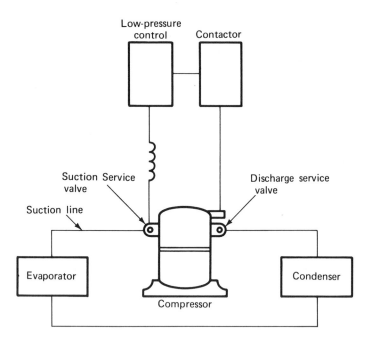

Fig. 5-1. Low pressure control in refrigerant system.

Table 5-1.   Approximate pressure control settings.

	VACUUM: ITALIC FIGURES		GAGE PRESSURE: LIGHT FIGURES			
	Refrigerant					
	12		22		502	
Application	Out	In	Out	In	Out	In
Ice cube maker, dry type coil	4	17	16	37	22	45
Sweet water bath, soda fountain	21	29	43	56	52	66
Beer, water, milk cooler, wet type	19	29	40	56	48	66
Ice cream, hardening rooms	2	15	13	34	18	41
Eutectic plates, ice cream truck	1	4	11	16	16	22
Walk-in, defrost cycle	14	34	32	64	40	75
Vegetable display, defrost cycle	13	35	30	66	38	77
Vegetable display case, open type	16	42	35	77	44	89
Beverage cooler, blower, dry type	15	34	34	64	42	75
Retail florist, blower coil	28	42	55	77	65	89
Meat display case, defrost cycle	17	35	37	66	45	77
Meat display case, open type	11	27	27	53	35	63
Dairy case, open type	10	35	26	66	33	77
Frozen food, open type	−7	5	4	17	8	24
Frozen food, open type, thermostat	$2°F$	$10°F$	—	—	—	—
Frozen food, closed type	1	8	11	22	16	29

The high pressure control is connected so that it stops the compressor when the head pressure exceeds the control settings. These high pressures could be caused by insufficient condenser cooling, excessive refrigerant charge, air in the refrigerant lines, or other abnormal conditions. The high pressure control is connected to the high side of the system, preferably where the compressor discharge valve cannot be back seated far enough to render the control inoperative (Fig. 5-2).

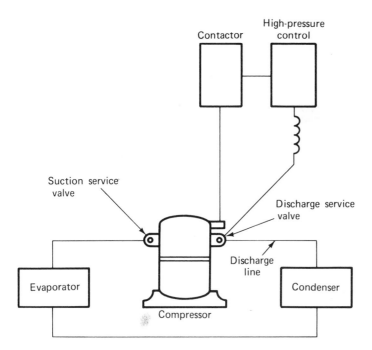

When both the high and low pressure controls are embodied in a single control the device is called a *high-low pressure control* or *dual pressure control*. This control provides savings in space, original cost, and wiring.

Pressure control electrical ratings have a range of about 24 full load amperes and 102 locked rotor amperes on 120V or 240V. See Fig. 5-3 for typical wiring diagrams.

Oil failure controls are designed to provide compressor protection by guarding against low lube oil pressure (Fig. 5-4).

A built-in time delay switch allows for oil pressure pickup on starting and avoids nuisance shutdowns on oil pressure drops of short duration during the running cycle.

In operation the total oil pressure is the combination of crankcase pressure and the pressure generated by the oil pump. Net oil pressure available to circulate oil is the difference between total oil pressure and refrigerant pressure in the crankcase; total oil pressure – refrigerant pressure = net oil pressure. This control measures that difference in pressure, referred to as *net oil pressure.*

When the compressor starts, a time delay switch is energized. If the net oil pressure does not increase to the control cut-in point within

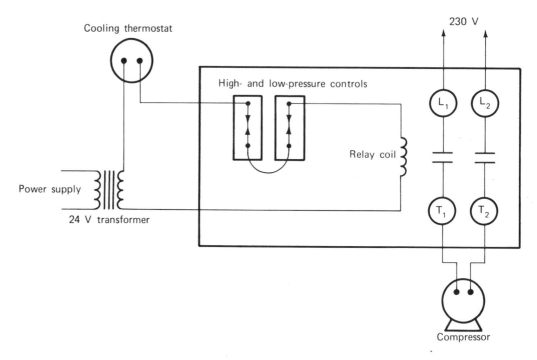

Fig. 5-3. Typical electrical diagram of high and low pres-
    sure controls.

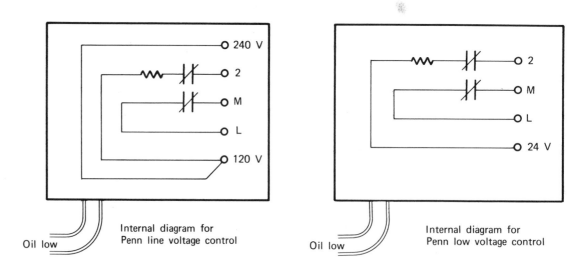

Fig. 5-4. Internal diagram for Penn line voltage and low
    voltage controls.

the required time limit, the time delay switch trips to stop the compressor. If the net oil pressure rises to the cut-in point within the
required time after the compressor starts, the time delay switch is
automatically de-energized and the compressor continues to operate
normally.

If the net oil pressure should fall below the cut-out setting during
the running cycle, the time delay switch energizes and, unless the net

oil pressure returns to the cut-in point within the time delay period, the compressor stops.

The time delay switch is a trip-free thermal expansion device. The time delay unit is compensated to minimize the effect of ambient temperatures from 32° F to 150° F. Timing is also affected by voltage variations. The manufacturer's recommendation should be followed during installation and adjusting. Normally, connect the oil pressure line to the pressure connector labeled OIL and the crankcase line to the pressure connector labeled LOW. Wire as suggested for specific equipment (see Fig. 5-5).

Oil failure controls are electrically rated for pilot duty only.

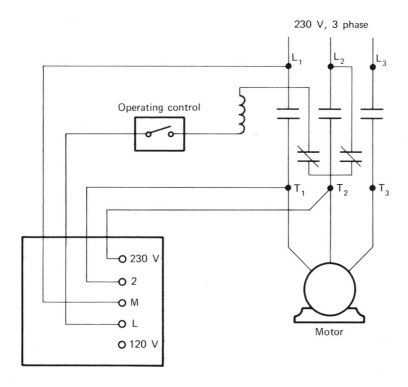

Fig. 5-5. *Penn Control wiring diagram used on 230 V system.*

QUIZ 5

1. The purpose of the high pressure control is to _____ .

2. The high pressure control is connected to the _____ side.

3. The low pressure control may be used as a _____ device in refrigeration work.

4. The oil failure control works on _____ oil pressure.

5. Net oil pressure is the_____ oil pressure and refrigerant pressure.

6. The oil failure control may be used only as _____ electrically.

7. Nuisance shutdowns with the oil failure control are prevented by the _____ .

8. Pressure controls may be either _____ or _____ reset.

9. The main purpose of low pressure controls in air conditioning is to _____.

TEMPERATURE CONTROLS, HUMIDISTATS, AND AIRSTATS

Any home, regardless of location, age, or architecture, can now provide its owner with year-round comfort and safety with a well designed, properly installed, and carefully controlled air conditioning system.

Maintaining perishable foods at the necessary temperature would also be very difficult, at best, if it were not for the development of modern temperature controls.

DEFINITIONS

Temperature controls are sensing devices used to maintain a desired temperature by switching on or off the necessary equipment.

Humidistats control the humidity within a structure by controlling the operation of a humidifier.

Airstats are safety devices used to control ventilation fans during abnormal conditions.

OPERATION

Refrigeration temperature controls are used to maintain the desired temperature within a refrigerated commercial case or a domestic refrigerator.

In the commercial case the controller is used to actuate a motor contactor to start or stop the compressor motor on temperature demand (Fig. 6-1).

Refrigeration controls are manufactured with the temperature sensing device contained within temperature sensitive power elements (Fig. 6-2).

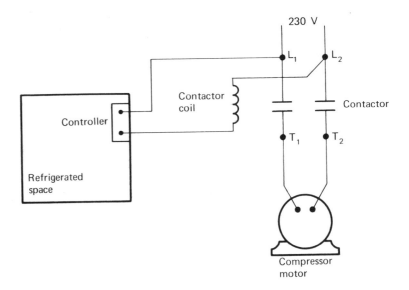

Fig. 6-1. Wiring diagram for refrigeration temperature control.

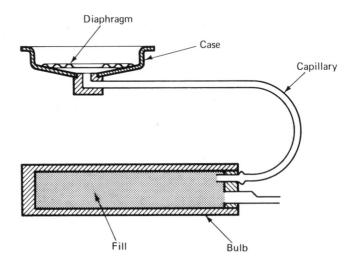

Fig. 6-2. Liquid filled remote bulb controller.

The vaporization or vapor-pressure element is based on the principle that the boiling temperature of a liquid depends upon the vapor pressure at the liquid surface. By partially filling a bulb with liquid and connecting the vapor space to a pressure sensitive element a closed system is formed in which the vapor pressure will depend upon the temperature of the bulb. In this type of element the pressure sensitive mechanism is usually a bellows (Fig. 6-2).

Different applications of these controls demand different types of charges within the power elements, much the same as the different charges in thermostatic expansion valve power elements. These different charges are vapor, liquid, and cross-ambient.

Another temperature control with which we are more familiar is the *room thermostat* that controls the air conditioning temperature in our homes.

In its simplest form, a thermostat is a device which responds to air temperature changes and causes a set of electrical contacts to open or close. This is the basic function of a thermostat, but there are many different types designed to perform a variety of switching functions.

One of the early types of heating systems that was capable of some degree of automatic control was the hand-fired coal furnace. Thermostatic control of this system was accomplished with a SPDT thermostat and damper motor. This was a long way from the completely automatic systems of today, but it was the beginning of automatic control for residential heating systems.

The heart of both this early system and the most modern systems of today is, of course, the room thermostat. Therefore, we shall begin with the bimetal room thermostat.

The bimetal thermostat gets its name from the fact that it uses a bimetal to open or close a set of contacts upon an increase or decrease in room air temperature.

A bimetal is made of two pieces of metal, which, at a given temperature, are the same length. If we increase the temperature of these two pieces of metal, one will become longer than the other. This is because they are different metals having different rates of expansion. These two metals are welded together in such a way that they become one solid piece, but they still keep their individual characteristics of different rates of expansion (Fig. 6-3).

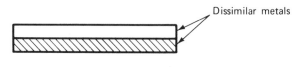

Fig. 6-3. Bimetal.

When we apply heat, one piece expands at a faster rate than the other piece. In order for one piece to become longer than the other, it must bend the entire bimetal into an arc (Fig. 6-4).

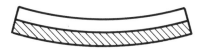

Fig. 6-4. Bimetal in heated condition.

If we now anchor one end of the bimetal to something solid, the free end will move down or up with an increase or decrease in temperature. By attaching contacts to the free end and placing stationary contacts nearby, we can get different switching actions with changes in temperature (Fig. 6-5).

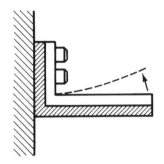

Fig. 6-5. Anchored bimetal.

The first bimetal room thermostats produced unsatisfactory results due to the unstable action of the contacts. Due to the relatively small differences in room air temperature the bimetal could not develop enough contact pressure to obtain a positive electrical connection.

With the development of the permanent magnet it was possible to obtain the convenience of a control system incorporating the best features of modern control circuits (Fig. 6-6).

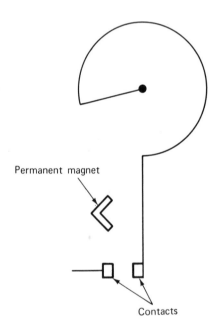

Permanent magnet

Contacts

Fig. 6-6. Thermostat-type bimetal.

SNAP ACTION VERSUS MERCURY SWITCH

Thermostats are available with either snap action or mercury switches. A description of each type follows.

Snap action switches are constructed with a fixed contact securely attached to the base of the thermostat. This contact is mounted inside a round permanent magnet which provides a magnetic field in the area of the contact. The movable contact is attached to the bimetal and upon a

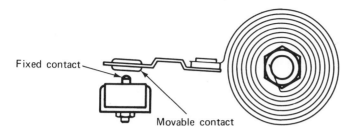

Fig. 6-7. Typical thermostat bimetal.

decrease in temperature (on heating models) moves slowly toward the fixed contact (Fig. 6-7).

As the movable contact enters the magnetic field around the fixed contact, the magnetic field pulls the movable contact against the fixed contact with a positive "snap." Because the movable contact has a floating action, it closes with a clean snap, without contact bounce. This floating action also eliminates any tendency for the contacts to "walk" while opening.

As the bimetal becomes warmer (on a heating model) it wants to pull the movable contact away from the fixed contact. Because the movable contact is in the magnetic field surrounding the fixed contact, the bimetal does not—at this instant—have enough force to overcome the magnetic field. As the bimetal continues to warm up and bend, it soon develops enough force to overcome the magnetic field and the movable contact breaks away with a positive snap. When we use SPDT switch action, another fixed contact is used, located so that when one contact is broken another is made (Fig. 6-8).

All snap action thermostats are supplied with a dust cover to prevent dirt and other contaminants from getting on the contacts. Should it become necessary to clean these contacts, never use a file or sandpaper. A clean business card or smooth cardboard should be inserted between the contacts. With gentle pressure on the movable contact, pull the card back and forth to clean the dirt or film from the contacts.

Mercury switches perform the same switching function as snap action, but the switching action is accomplished by a globule of mercury moving between two or three fixed probes sealed inside a glass

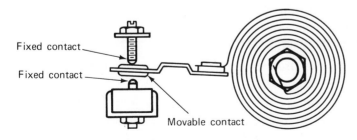

Fig. 6-8. Single pole-double throw thermostat.

tube. Two probes are used on SPST switches (Fig. 6-9) while three probes are used on SPDT models (Fig. 6-10).

These mercury tubes are attached to the thermostat bimetal and perform the desired function.

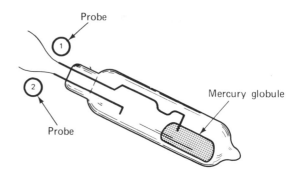

Fig. 6-9. SPST mercury switch.

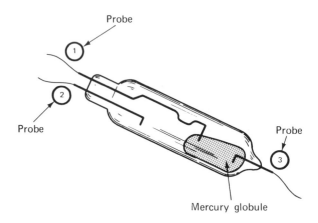

Fig. 6-10. SPDT mercury switch.

FORCED WARM AIR SYSTEM: NONANTICIPATED THERMOSTAT

If the temperature lever on the thermostat is set at 75° F and the furnace has been off for some time, the temperature in the room drops slowly. The bimetal element has been following the air temperature change and closes the electrical contacts at 75° F. This causes the heating system to start. At this moment no warm air is being delivered to the room because the heating system must warm up to the "Fan on" setting on the fan control.

While the heating system is warming up, the room air temperature will continue to drop slowly. Depending on the type of heating system, the room air temperature will drop to 74.5° F, 74° F or below before the blower comes on and the warm air is "felt" by the thermostat bimetal. This difference in temperature between the point at which the thermostat contacts close and the point the air temperature at the thermostat starts to rise is known as *system lag*. The amount of system lag in degrees Farenheit will depend on the thermostat location, type, and size of the furnace, and the design of the air distribution system.

With the furnace on and the blower running, the room air temperature will continue to rise. If the thermostat has a mechanical differential of 2° F, the electrical contacts will open at 77° F and shut down the primary control on the furnace. However, the furnace is still hot and the blower will continue to deliver warm air to the room until the furnace temperature drops to the "Fan off" setting on the fan control. The additional heat that has been delivered to the room after the thermostat contacts have opened is called *overshoot*. This overshoot can carry the room air temperature to 77.5° F, 78° F, or higher.

FORCED WARM AIR SYSTEM: ANTICIPATED THERMOSTAT

To reduce the wide differential resulting from a nonanticipated thermostat, we simply add a small amount of heat to the bimetal element so that it is slightly warmer than the surrounding room air temperature. We do this by placing a resistor in the thermostat close to the bimetal element. This resistor is in series with the contacts. When the contacts close and the primary control is energized, the current flowing through the primary control must also flow through the resistor. The current flowing through the resistor causes it to heat up; which in turn heats the bimetal element. Thus, we "anticipate" the point at which the thermostat contacts should open to give us a narrow differential.

Because the furnace has been on a shorter period of time, there is less heat left in the heat exchanger. This means we will have less overshoot. In the meantime, the bimetal element cools down because it is no longer being heated by the anticipator. Thus, we begin the cycle all over again.

We have not eliminated system lag or overshoot, but through the use of heat anticipation we reduce these factors to the negligible point. Through proper system design and thermostat location, it is not unusual to obtain room air temperature differential no greater than ½° F by the proper use of heat anticipation.

TYPES OF HEAT ANTICIPATORS

Heat anticipators are made in two types, fixed and adjustable. A fixed anticipator can be either wire wound or the carbon resistor type. Earlier models of low voltage thermostats used wire wound fixed anticipators. They were dipped in an insulating material and color coded to indicate the primary control with which they were to be used.

Present day thermostats with fixed anticipators use tubular resistors, which is also color coded to indicate the current draw of the primary control. These are supplied either as "nonremovable" (riveted in place) or "removable" (attached with a screw). Fixed anticipators must match the current draw of the primary control. A number of different ranges are available (Fig. 6-11).

The most versatile heat anticipator is, of course, the adjustable type (Fig. 6-12).

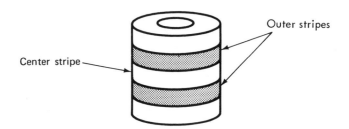

Fig. 6-11. Fixed heat anticipator.

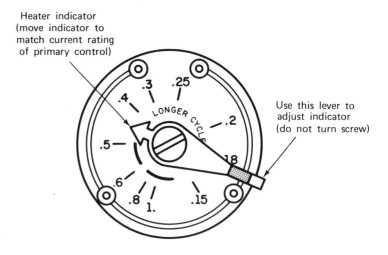

Fig. 6-12. Adjustable heat anticipator.

HOW TO SET AN ADJUSTABLE ANTICIPATOR

The primary purpose of the adjustable anticipator thermostat is to provide a single thermostat to match almost any type of primary control in the field. Before installing the thermostat, check the nameplate on the primary control. Make sure that the primary control is a low voltage model (25 V), not line voltage! A low voltage thermostat will burn out immediately if it is connected to a line voltage primary control! If, for example, the nameplate tells you that it is a low voltage control and the current is marked .45 amp, set the indicator (Fig. 6-12) on the adjustable anticipator to .45 amp. By matching the adjustable anticipator to the current rating on the primary control you are assured of the best possible heat anticipation for the system.

This anticipator is a .25 watt carbon-type resistor similar to that found in a radio or television set. This type of anticipation is known as *off-cycle anticipation*. To understand this type of anticipation better, let's look first at how it is used in a typical cooling system (Fig. 6-13).

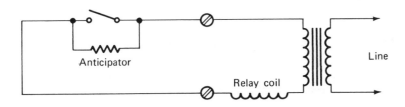

Fig. 6-13. *Off-cycle anticipation schematic diagram.*

In a cooling system heat is added to the thermostat bimetal during the off cycle. This is just the opposite of heat anticipation in a heating system. The anticipator in a cooling system is in parallel with the contacts of the thermostat. When the contacts close, this provides a low resistance path for the current in the thermostat circuit, which pulls in the relay on the air conditioning system. When the thermostat is satisfied, the contacts open. We now have a high resistance path from the transformer, through the cooling anticipator and the winding of the relay coil, back to the transformer. Because we have a high resistance in the cooling anticipator, it drops the voltage to the point that the relay will not pull in.

The current flowing through the cooling anticipator heats up the thermostat bimetal. This causes the bimetal to be warmer than the surrounding room air temperature which is also rising (the system is off). This "false" heat causes the contacts to close before the room air temperature reaches the cut-in point. Thus, we are bringing the cooling system on sooner, reducing "system lag" to a minimum and providing a narrow differential for the cooling system. Cooling anticipators are provided on all low voltage thermostats, both heating and cooling models.

In recent years there has been an increased demand for greater comfort and efficiency in indoor comfort systems. The staging thermostat has been designed to meet these needs. (Fig. 6-14).

The staging thermostat is used on these systems which are designed for one or two stages of heating or one or two stages of cooling or any combination of heating and cooling.

In a typical system of two stage heating and two stage cooling, when the thermostat is in the heating position, the heating system

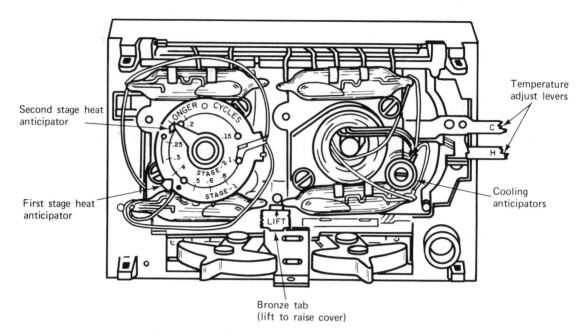

Fig. 6-14. Multi-stage thermostat.

operates at a reduced BTU input level during mild weather. As the weather becomes colder, this reduced level of input is not sufficient to maintain the desired comfort level and the thermostat automatically brings on an additional level of input to the heating system. In the cooling position the situation would be similar except that we would be bringing on one or two levels of cooling instead of heating. These thermostats are also available with an automatic changeover. In the automatic position all that is necessary is to set the desired level of heating and cooling, and the thermostat will automatically change the system from heating to cooling and back to heating, based on the temperature settings on the thermostat. There are two separate temperature levels on these thermostats, one marked *C* for cooling and the other marked *H* for heating.

FAN SWITCHES

The fan switch on a thermostat has two positions, one for automatic operation and one for continuous operation.

When the switch is set in the "auto" position the fan will operate only on demand from the part of the unit that is in use at the time (heating or cooling).

If the switch is positioned at the "on" side the fan will operate continuously regardless of the system demand. The fan will also operate in this position when the system switch (HEAT or COOL) is turned to the off position.

The circuit wiring with staging thermostats will, of course, vary from installation to installation due to the number of possible systems that can be used. We shall use a two stage heating-two stage cooling wiring diagram to review thermostat function. From the diagram below (Fig. 6-15) we can see that the system switch is on "heat" and the fan switch is on "auto." From one side of the transformer the circuit is to terminal *RC,* through the jumper wire to *RH,* through the internal wiring to the bar contact on "heat" and on to the stage one and two heat anticipators. You will note that we also have a circuit to terminal

CIRCUIT WIRING

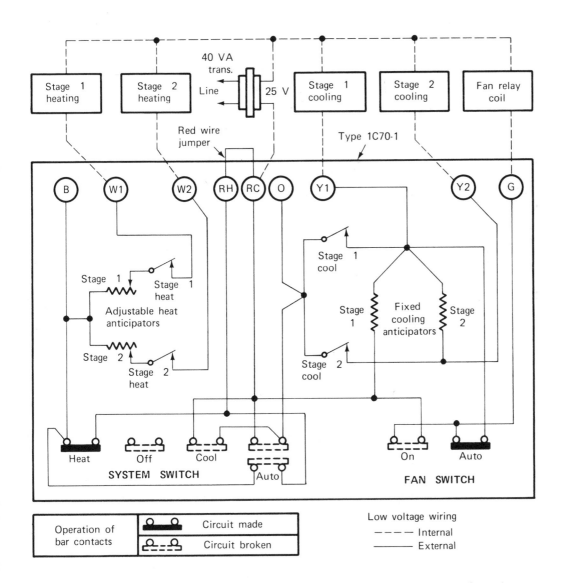

Fig. 6-15. Multi-stage system wiring schematic.

B which will energize another circuit continuously. Upon a call for heat, the stage one heating switch will close, giving us a circuit to *W1*, then to the first stage of the heating system, and back to the other side of the transformer. If the temperature continues to drop, the stage two heating switch will close, giving a circuit through *W2*, to the second stage of the heating system, and back to the other side of the transformer.

When the selector switch is placed in the "cool" position we break the "heat" bar contact and make the "cool" contact. This gives us a circuit from *RC* to the "cool" contact, through the internal jumper, to the "auto" contact, and on to stage one and stage two contacts. When stage one contact closes we have a circuit through *Y1*, stage one cooling, and back to the other side of the transformer. If this is not sufficient cooling, stage two contacts close and this circuit is made through *Y2*, stage two cooling, and back to the other side of the transformer.

MODULATING THERMOSTATS

Modulating control systems are built around the Wheatstone Bridge principle. Both the controller and the controlled apparatus, usually a modulating motor operating a damper or a water valve, use this principle. Both the thermostat (controller) and the motor have potentiometers in them.

The operation of this type of circuit is as follows (Fig. 6-16). If the temperature at the controller rises:

1. The wiper on the controller potentiometer moves toward the *W* terminal reducing resistance between *W* and *R* at the controller.
2. Current flow from the transformer through the *W* terminal of the controller is increased, and the increased current in the corresponding relay coil pulls in the DPST relay switch to contact 1.
3. Current from the transformer then flows through the relay switch to 1 on the motor.
4. The motor runs counterclockwise to reposition the controlled device. As the motor runs it moves the wiper on the motor potentiometer toward the *G* terminal.
5. When the wiper on the motor potentiometer reaches the point where the resistance between *T* and *G* on the motor potentiometer equals the resistance between *R* and *W* on the controller potentiometer, the current in the relay coil is equalized.
6. The relay contacts break, stopping the motor and the circuit is again in balance.

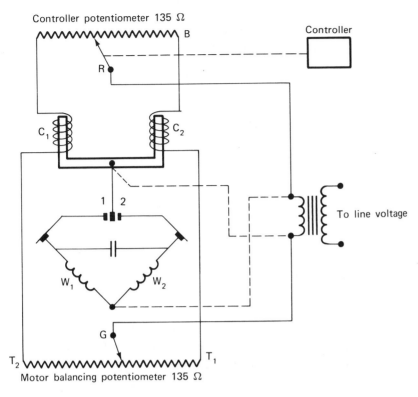

Fig. 6-16. Motor balancing potentiometer (135 ohms).

On a temperature drop, the current flows to the other side of the relay and the motor runs in the opposite direction.

The potentiometer wiper is moved by a gas filled bellows through a series of levers and springs.

THERMOSTAT LOCATION

We have mentioned earlier that thermostat location is very important for the successful performance of the total system.

For your guidance, we offer the following suggestions for proper location of the thermostat.

1. Always locate the thermostat on an inside wall. If it is located on an outside wall, overheating during cold weather is likely to occur because the thermostat will always "feel" the cold.
2. Avoid false sources of heat such as lamps, television sets, warm air ducts, or hot water pipes in the wall, and locations where heat producing appliances like ranges, ovens, dryers, and so on are located on the opposite side of the wall. Locations near windows may cause direct sunlight to reach the thermostat.

3. Avoid sources of vibration like sliding doors, closet doors, and room doors. Always locate the thermostat at least four feet from such sources of vibration and near a wall support if possible.

THERMOSTAT VOLTAGE

Thermostats are available in all common voltages and must be used with the stated voltage. If used otherwise, they will be permanently damaged or improper operation will result.

HUMIDISTATS

Humidistats are used to control humidification equipment on air conditioning systems during the heating cycle. It is desirable to add humidity to some structures because the air becomes drier when heated by the heating equipment. There are several ways to wire these controls into a system. However, they should be wired so that the humidifier operates only when the fan is in operation to prevent moisture from collecting on the heat exchanger during the off cycle. One suggested wiring diagram is shown in Fig. 6-17. This diagram provides the safety of not allowing the water solenoid to be actuated unless both sources of electricity are available.

Humidistats may be used on either line or low voltage. They have a moisture-sensitive nylon ribbon which is wound around three bobbins to give, effectively, four element control. Positive on-off settings are provided for manual operation. The switch is a SPST, snap action switch.

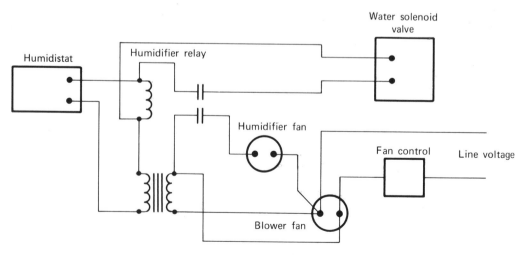

Fig. 6-17. Typical humidifier wiring diagram.

The airstat controller is used as a fan safety cut-off. It stops fan operation whenever the return air plenum temperature rises to a point that indicates the possibility of a fire. The controller locks out to prevent the premature return of a fan operation. The switch may be either bimetal actuated or a fusible link operating a SPST type of control. When this control is actuated it must either be manually reset or the fusible link replaced.

AIRSTATS

To wire this control into the electrical system see the following diagram (Fig. 6-18).

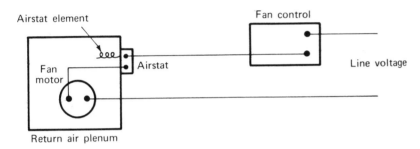

Fig. 6-18. Airstat wiring diagram.

QUIZ 6

1. Where is a refrigeration temperature control most likely to be used?

2. The temperature sensing element on a temperature control is similar to the power element of an _____.

3. Another type of temperature control is the _____ _____.

4. The part of the thermostat to which the movable contact is attached is called a _____ .

5. The fixed contact on a thermostat is mounted inside a _____ .

6. Thermostat contacts should be cleaned with a _____ .

7. The anticipator in a heating thermostat causes the thermostat to (open, close) before the desired temperature is reached.

8. New types of anticipators are of the (adjustable, carbon resistor) construction.

9. The voltage (does, does not) matter on a thermostat.

10. The control circuit which causes the cooling unit to come on (is, is not) made *through* the cooling anticipator.

11. Multiple units are started, one at a time, by a _____ _____ .

12. The fan will operate continuously in the _____ position.

13. Modulating thermostats are designed on the _____ principle.

14. The thermostat should be located on an _____ wall.

15. Airstats are used as _____ devices.

16. Airstats are mounted in the _____ air plenum.

17. A humidistat is used to _____ .

18. Humidistats may be used on _____ _____ voltage.

19. Humidistats sense the moisture content of the air with a _____ .

20. Airstats are actuated by either a _____ or _____ .

7

POWERPILE AND
THERMOCOUPLE SYSTEMS

A commonly used source of electric current in which heat is transformed into electric energy is the thermocouple. In modern day heating systems, the thermocouple has become a widely used safety device. It is normally used in conjunction with pilot safety controls. (Discussed in a later chapter). The basic operation of the thermocouple must, therefore, be understood by the service technician.

A *thermocouple* is a junction of two bars, wires, or similar devices of dissimilar metals that produces electric current when heated.

A *powerpile* consists of a series of thermocouples connected to produce a higher current than a single thermocouple.

DEFINITIONS

When two pieces of dissimilar metals, usually iron and copper, are connected together and heated, an electric current will flow. This current is caused by the interaction of the two metals and the resulting movement of electrons (Fig. 7-1).

The current produced between the hot and cold junctions will increase with an increase in temperature, up to a given current rating. Above this level there will be little, if any, rise in current regardless of the amount of heat applied. These devices should not be overheated because excessive temperatures will result in a burned out junction and a ruined thermocouple.

OPERATION

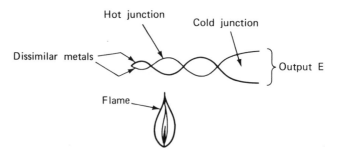

Fig. 7-1. Thermocouple.

Thermocouples are normally used as safety devices and will hold an electromagnetic relay in the "pulled in" position after the armature is manually pushed into the electromagnetic coil. If, for some reason, the current is reduced or stopped, the armature will be pulled out of the coil by spring pressure.

Thermocouples are rated at 30 millivolts (mV) when the end is inserted into the pilot flame from 3/8 to 1/2 in. (Fig. 7-2).

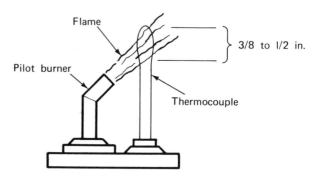

Fig. 7-2. Relationship of pilot flame and thermocouple.

These devices are connected directly to the individual controls they operate. Thermocouples do not depend on any outside voltage to perform and should not be connected into any electrical control circuit with voltage different from their output (Fig. 7-3). A damaged control will result.

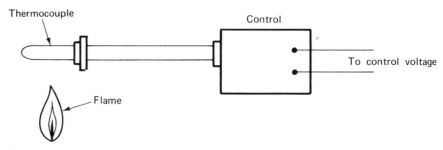

Fig. 7-3. Thermocouple regulated control.

Powerpile systems produce much higher current than do thermocouples. Powerpiles will produce from 250 mV to 750 mV, depending on the number of thermocouples used.

As stated in the definition, a powerpile is simply a series of thermocouples and will produce current in proportion to the number of these and the heat applied (Fig. 7-4).

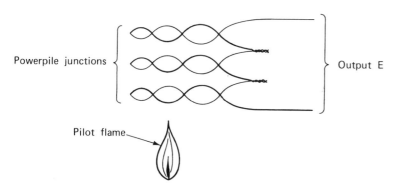

Fig. 7-4. Fundamental powerpile element.

Powerpile systems are usually used to produce the full operating current for a millivolt control circuit. See Fig. 7-5 for a wiring diagram.

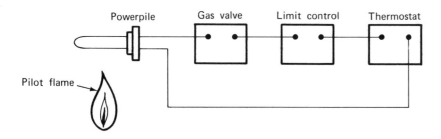

Fig. 7-5. Typical millivoltage wiring diagram.

As with a single thermocouple, this kind of system must be separated from any other electrical current.

For proper operation of the heating equipment, the proper thermostat must be used, or one with an adjustable heat anticipator. All electrical connections should be scraped and soldered to reduce voltage drop to a minimum.

Caution: do not use a 30 mV control with powerpile voltage because a burn out of the control will result, similar to using 115 V controls on a 230 V source.

QUIZ 7

1. A thermocouple is made of _____ .

2. Thermocouple output is _____ with an increase in temperature.

3. Millivolt and 24 V systems (can, cannot) be mixed.

4. A powerpile is _____ .

5. A powerpile produces from _____ to _____ mV.

6. Thermocouples are normally used as _____ .

7. The output of a thermocouple is _____.

8. A thermocouple (will, will not) pull in an armature alone.

9. A thermocouple will operate best when inserted from _____ in. to _____ in.

10. Thermocouple output is _____ than a powerpile.

8

GAS VALVES AND PILOT SAFETY CONTROLS

Automatic control of gas to the burners of a heating device is accomplished by use of electrically powered main line gas valves. Their safe operation is assured through the use of pilot safety controls, in conjunction with a thermocouple. Automatic operation is accomplished by use of a thermostat.

DEFINITIONS

Gas valves are normally closed solenoid valves designed for the control of gas to the combustion equipment.

Pilot safety controls provide safe, automatic shut-off of gas valves during pilot flame failure in appliances.

VOLTAGE

Gas valves are manufactured in a wide variety of voltages. The type used on any particular installation must correspond to the control voltage. Voltage ratings range from millivolts (250-750) to 24 V, 115 V, and 230 V. Some are manufactured with dual voltage coils.

Pilot safety controls are manufactured with solenoid coils for millivolts *only*. These coils may be one of two types: (1) manual reset, where the armature must be pushed into the coil manually, and (2) automatic reset, where the electromagnetic coil has enough power to pull in the armature.

OPERATION   Gas valves are normally closed. When the thermostat or switch directs electricity to the solenoid, it pulls the plunger into the electromagnetic coil and lifts the attached valve seat disk (Fig. 8-1). Gas then flows through the main valve port until the electrical circuit is interrupted. This releases the plunger. The weight of the plunger, the seat assembly, and the gas pressure on top of the valve seat insure a tight shut-off.

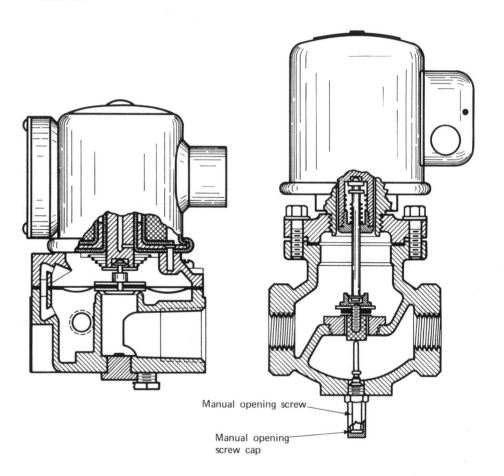

Manual opening screw

Manual opening screw cap

Fig. 8-1. Cross sectional views of two types of gas valves.

Another type of gas valve is the combination gas valve. This valve provides all manual and automatic control functions required for operation of gas fired heating equipment.

This valve includes (1) a gas operated diaphragm main valve, (2) a permanently lubricated shear seal, disk-type gas cock, (3) a electromagnetic safety valve, (4) a pilot gas adjustment valve and pilot line filter, and (5) a pressure regulator (Fig. 8-2).

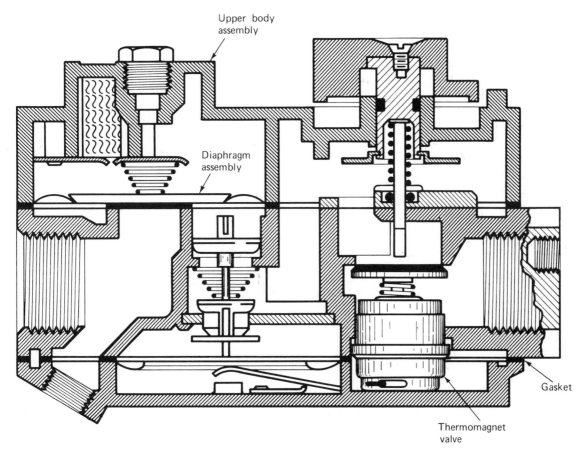

Upper body
assembly

Diaphragm
assembly

Gasket

Thermomagnet
valve

Fig. 8-2. Cutaway view of a combination gas valve.

The main line gas valve opens and closes in response to a thermo-stat and/or limit control. The main line valve may also function simul-taneously as a pressure regulator. This type of valve is called a combination gas valve (Fig. 8-3).

Pilot safety controls are subdivided into two distinct types, (1) thermopilot valves and (2) thermopilot relays.

The thermopilot valve is a 100 per cent safety shut-off control for gas fired appliances and heating equipment. A push button provides manual reset operation (Fig. 8-4). This valve is installed in the main gas line ahead of the main gas valve (Fig. 8-5).

Current from a thermocouple energizes an electromagnet to hold the valve open after it is manually reset. Loss of current to the elec-tromagnet due to low pilot flame, pilot outage, or limit switch opera-tion causes the electromagnetic valve to snap closed, shutting off the gas flow. Depressing and holding the button opens the gas valve to allow pilot ignition. After the pilot burns about 60 seconds, electrical current from the thermocouple is sufficient to hold the valve open.

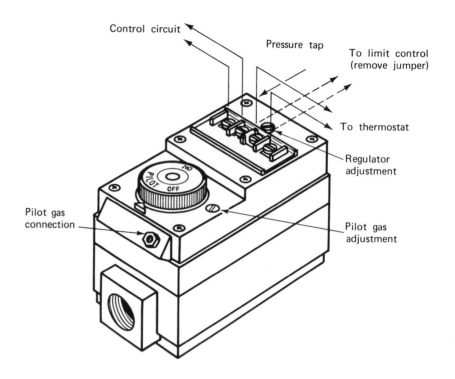

Fig. 8-3. Combination gas valve.

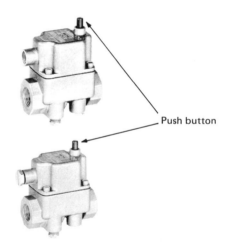

Fig. 8-4. Two types of thermopilot valves (Courtesy of ITT General Controls).

Thermopilot relays can also be subdivided into (1) automatic reset and (2) manual reset.

Automatic thermopilot relays provide safe, automatic shut-off of gas valves during pilot flame failure in appliances using line or low voltage operated gas valves. When the pilot flame is safely restored the relay automatically allows the gas valve to open. This prevents a false shutdown of the appliance when gas pressure varies or drafts affect the pilot flame. The relay also features switching for automatic ignition during the pilot flame failure (Fig. 8-7).

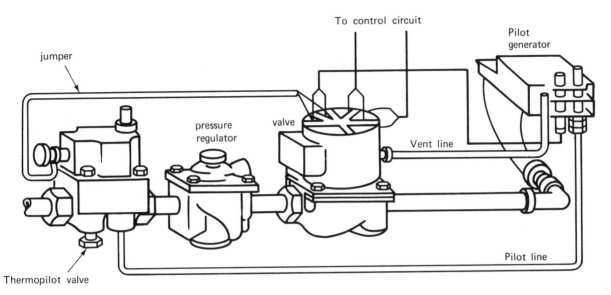

Fig. 8-5. Installation of thermopilot valve.

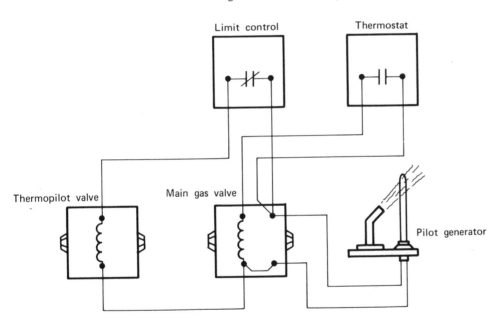

Fig. 8-6. Millivolt system with 100% shutoff wiring diagram.

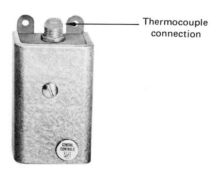

Fig. 8-7. Thermopilot relay (Courtesy of ITT General Controls).

The relay is powered by millivoltage from the pilot generator, which is heated by the pilot flame. This energizes the relay and closes its contacts to complete the electrical circuit to the main gas valve, allowing gas to flow (Fig. 8-8).

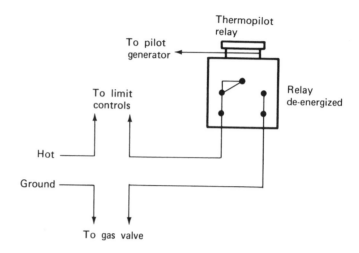

Fig. 8-8. Thermopilot relay wiring diagram.

If insufficient millivoltage is produced by the heat of the pilot flame, the relay is de-energized; the relay opens its contacts and the circuit to the main gas valve is broken. The main gas valve then shuts off the gas.

In single pole–double throw models, the de-energized relay also completes a second circuit which can be used for ignition or signal alarm during pilot flame failure (Fig. 8-9).

The manual reset thermopilot relay is an electric contacting device, which provides safe lighting, automatic safety shut-off for gas appliance controls, and a manual reset indicator.

A reset on all manual reset relays shows if the relay is on or off; this simplifies lighting operations and permits manual shut-off for the

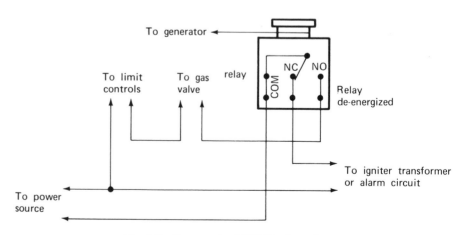

Fig. 8-9. De-energized SPDT relay schematic.

appliance. A pilot gas valve in a single unit provides safety shut-off control of the pilot gas in addition to main gas shut-off if the pilot is extinguished.

Electrical power for operating the electromagnetic relay is supplied by a single thermocouple (30 mV). On SPST models, turning the knob to pilot reset opens the relay contacts. If a pilot valve is also incorporated, gas also flows to the pilot. After the pilot has been burning for about 1 minute, the reset can be released and allowed to turn to ON (Fig. 8-10).

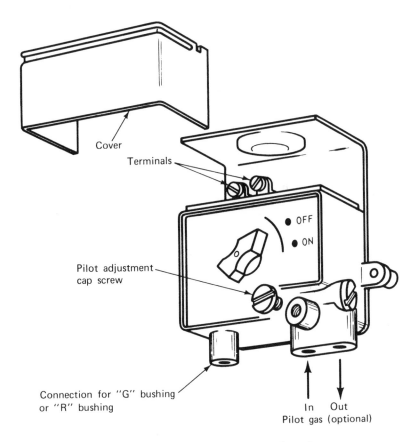

Cover

Terminals

● OFF

● ON

Pilot adjustment cap screw

Connection for "G" bushing
or "R" bushing

In   Out
Pilot gas (optional)

Fig. 8-10. Manual reset thermopilot relay.

If the pilot flame becomes unsafe, the relay contacts automatically open, the pilot valve closes, and the reset turns to OFF. The system can be shut off at any time by turning the reset to OFF.

HIGH-LOW FIRE VALVES

The high-low-off type of combination diaphragm valve provides all manual and automatic control functions required for operation of gas fired heating equipment. They have an internally vented diaphragm-type main control valve and a separate thermomagnetic safety valve

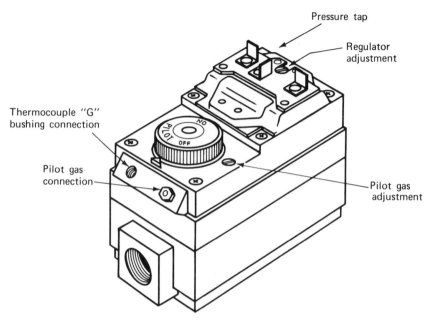

Fig. 8-11. High-low-off combination gas valve.

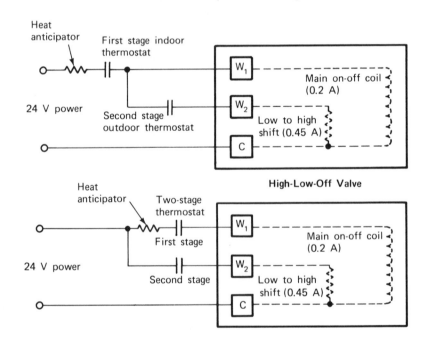

Fig. 8-12. Schematic diagram for high-low-off combination
gas valve.

with pilot gas adjustment and pilot filter. Most models include a
pressure regulator function for use with natural gas (Fig. 8-11).

The main line gas valve diaphragm opens and closes in response to
a thermostat and/or a limit control. The main line diaphragm valve
functions simultaneously as a pressure regulator on natural gas models.
The models used with LP gas are not equipped with a regulator.

When the electrical circuit is completed between C and W, terminals of the valve automatically open to the preset low fire position. Then, when the circuit is completed between C and W_2, the valve opens to the high fire position. The low fire to high fire shift is accomplished by a heat motor in the valve operator. Time must be allowed for this heating action to be accomplished (Fig. 8-12).

This type of valve is used when the full BTU rating of the appliance is not needed but when some heating is desired, such as in the spring and fall months or whenever mild weather exists.

QUIZ 8

1. The electrical power supplied to pilot safety control solenoid coils is _____ .

2. In what position is the gas valve when it is de-energized?

3. Gas pressure regulators are used on (natural, LP) gas valves?

4. Name two types of thermopilot valves.

5. The SPDT thermopilot relays can be used as _____ when the thermocouple is not heated enough.

6. The pilot gas valve, in a thermopilot relay, closes when the _____ goes out.

7. Turning the manual reset, on a thermopilot relay, to the off position _____ at any time.

8. The main burner flame burns at full rate when the _____ is complete between _____ and _____ on the high-low-off valve.

9. High-low-off valves are provided with pressure regulators for _____ gas.

10. Gas valves are _____ solenoid valves.

9

FAN CONTROLS
AND LIMIT CONTROLS

In the days of hand-fired coal furnaces and boilers, the person who did the shoveling waited around long enough to make sure the fire was burning properly. That's just one of the many jobs performed today by automatic controls. The "silent servants" are on duty around the clock to operate heating and cooling systems safely and economically and to provide comfortable living. Two more such "servants" are the fan control and limit control.

A *fan control* is a normally open, temperature actuated SPST electrical switch, which is mounted to sense the air temperature within the furnace.

A *limit control* is a normally closed, temperature actuated, SPST electrical switch, which is mounted to control the air temperature within the furnace.

Fan controls. Forced-circulation units have "heat watchers" in addition to the burning controls. These additional controls permit the circulating fan to operate only when there is enough heat available.

There are two types of fan controls: (1) temperature actuated (Fig. 9-1) and (2) electrically operated (Fig. 9-2).

The temperature actuated fan control makes use of a bimetal strip. This bimetal strip is inserted into the warm air plenum or directly into the air passage of the furnace.

Fig. 9-1. Heat actuated fan control (Courtesy of Honeywell. Inc.).

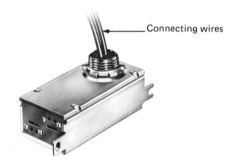

Fig. 9-2. Electrically operated fan control (Courtesy of Honeywell, Inc.).

As the temperature of the furnace increases the bimetal strip bends. When the fan's ON setting of the control is reached, usually 150°F, the snap action switch closes and completes the line voltage circuit to the fan motor (Fig. 9-3).

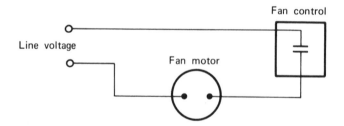

Fig. 9-3. Schematic of fan motor wiring.

When the thermostat is satisfied, or the limit control opens, the main gas valve closes, stopping the flow of gas to the burners. As the furnace cools down, the air delivered through it also cools down and causes the bimetal strip to return to its original position. When this happens the switch is opened and the fan motor stops. The fan turn-off temperature is usually about 100°F.

The electrically operated fan control (Fig. 9-2) provides timed fan operation for forced warm air furnaces when the heater coil is wired in parallel to the low voltage gas valve (Fig. 9-4). This control is particu-

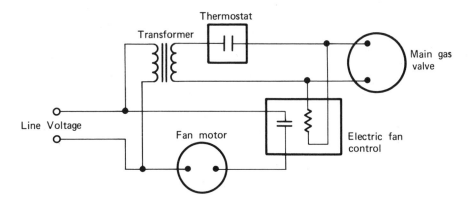

Fig. 9-4. Schematic diagram of an electrically operated fan control.

larly suited for counterflow and horizontal furnaces, because of the peculiarity of heat build up in the furnace. The heater coil uses 24 V ac. The switch is a SPST, heater actuated, bimetal control. The fan is usually started about one minute after the thermostat demands heat; the fan is stopped about two minutes after the thermostat is satisfied.

LIMIT CONTROLS

The limit control is used with all types of warm air furnaces to prevent excessive plenum temperatures and a possible resulting fire. The snap action switch is actuated by a flat bimetal element which is inserted into the plenum (Fig. 9-5).

Fig. 9-5. Limit control (Courtesy of Honeywell, Inc.).

The temperature range is adjustable from about 180° F to 250° F with a fixed differential of 25° F. The fixed setting is the point at which the SPST switch closes to bring the furnace back into operation. When the OFF setting, usually around 200° F, is reached, the electrical circuit is opened to the main gas valve, thus stopping the gas flow to the main burners. As the furnace cools down the bimetal element also cools, and when the predetermined OFF setting is reached the electrical circuit is made to the gas valve and operation resumes (Fig. 9-6).

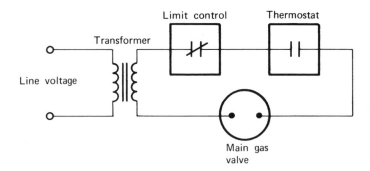

Fig. 9-6. Schematic diagram of limit control in the circuit.

The only time this control is designed to function is when there is insufficient air flow through the furnace. This control is usually designed for pilot duty only and will not accommodate heavy current through the contacts.

FAN-LIMIT CONTROL

Fan-limit (combination) controls provide fan and limit control on warm air heaters and furnaces. They are manufactured with a snap action, sealed switch. The contacts are designed for millivolt, 24V, 120V, or 240V circuits.

The limit switch breaks a circuit on temperature rise to stop burner operation if the plenum temperature reaches the indicated scale setting; this is similar to a single limit control.

On a temperature rise, the fan switch makes a circuit to start blower operation when the plenum temperature rises to the fan ON scale setting. The blower will stop when the plenum temperature drops to the fan OFF scale setting. This is identical in operation to individual fan control operation.

The combination fan and limit control combines the functions of the individual fan and limit controls into a single compact unit. The scale setting is simplified. The limit action cannot be set below the fan control action. A summer fan switch, on some models, is readily accessible without removing the cover and provides for the selection of continuous fan operation or automatic fan operation (Fig. 9-7).

For some suggested wiring diagrams see Figs. 9-8 and 9-9.

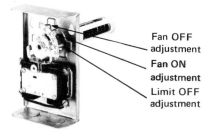

Fig. 9-7. Combination fan and limit control (Courtesy of ITT General Controls).

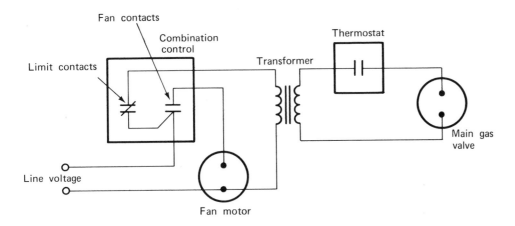

Fig. 9-8. Typical control circuit using combination fan and limit control.

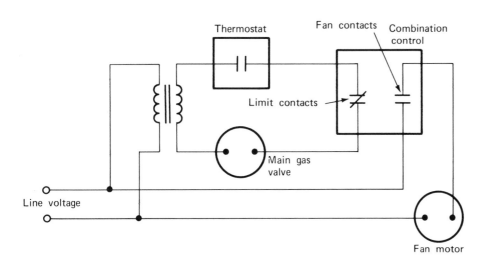

Fig. 9-9. Typical control circuit with fan control in the line voltage circuit and limit control in low voltage circuit.

QUIZ 9

1. The fan control directs electricity to the _____ on a rise in _____ .

2. The limit control directs electricity to the _____ on a _____ in temperature.

3. Two types of fan controls are _____ and_____ .

4. Both the fan and the limit control are actuated by a_____ element.

5. The _____ control can stop the flow of gas to the furnace the same as the _____ .

6. The limit control is basically a _____ device.

7. The fan control only operates when there is enough _____ to _____ the structure.

8. The _____ control setting cannot be set below the _____ control on the combination control.

9. The fan control is set to start fan operation at about _____ and stop fan operation at about _____.

10. The limit control is usually set to open the circuit at about _____.

AUTOMATIC
PILOT IGNITION

Provision must be made for proper ignition of the fuel and air mixtures supplied to the furnace. Gas fired burners are generally operated in an on and off mode, and some provision for automatic ignition is necessary. Gas burners usually employ a standing pilot. The pilot must be ignited manually; however, a hot wire system is sometimes provided for pilot ignition. Spark ignition for gas burners is becoming more common, especially for those installations where location of the burner makes manual pilot ignition difficult.

The *automatic pilot ignition* automatically and instantly lights or relights pilot gas.

DEFINITION

There are two main types of automatic pilot ignitors: 1) the most common type, the glow coil (Fig. 10-1), and 2) a newer type which ignites the pilot by a spark across a gap (Fig. 10-2).

OPERATION

The glow coil type lights a standing pilot. Incorporated in this unit is a purge time delay of approximately five minutes if the pilot flame goes out. Pilot burner relighting, when required, is automatic at all times that 25V ac power is supplied to the unit.

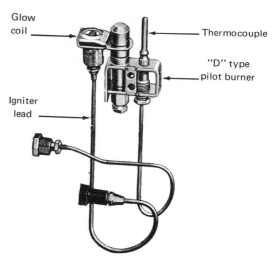

Fig. 10-1. Glow coil and pilot (Courtesy of Penn Controls, Division of Johnson Service Company).

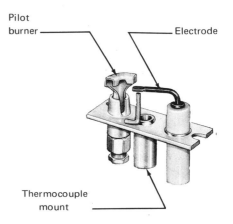

Fig. 10-2. Spark ignitor and pilot (Courtesy of Penn Controls, Division of Johnson Service Company).

NORMAL LIGHTING SEQUENCE (Fig. 10-3)

1. *A* and *B* valves are open (not shown). Only the electrical circuit is shown in this diagram.
   a. Pilot gas flows to pilot burner.
2. Power is available to the control.
   a. The warp switch is heated.
   b. The igniter is energized.
3. Pilot burner gas lights.
   a. The thermocouple is heated.
4. Pilot switch operates (usually about 20 sec delay).
   a. The normally closed contacts open; the normally open contacts close.
   b. The igniter is de-energized.

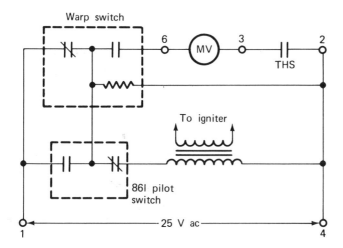

Fig. 10-3. Schematic wiring diagram.

5. The warp switch operates (approximately two minutes after Step 4).
   a. The thermostat and main valve circuit now have electrical potential.
6. The thermostat contacts close on a call for heat; the main gas valve opens.
   a. The main burner gas ignites and the system now operates.
   Note: Step 6 cannot occur unless Steps 4 and 5 are completed.

SAFETY SEQUENCES

No Gas On Start.

1. Warp switch and igniter are energized.
2. The warp switch operates (about two minutes later) to de-energize the igniter.
   a. The warp switch now cools.
3. When the warp switch cools (about five minutes) another ignition try is made. This will continue until the electrical circuit is interrupted.

Interruption of Gas During Running Cycle.

1. The pilot and main burner flames are extinguished.
   a. The gas supply may return without ignition.
2. The thermocouple cools in about 45 sec.
3. The pilot switch operates.
   a. The contacts return to the start position.
   b. The igniter and main gas valve is still de-energized.
   c. The warp switch heater cools.

4. After about five minutes the warp switch contacts reverse and Steps 2 through 6 of the normal lighting sequence are repeated. If the gas supply does not return Steps 1, 2 and 3 occur of *No Gas On Start.*

Power Interruption on Start.

1. Only the pilot gas flows.
2. Operating sequence will resume when the power is restored.

Power Interruption During Running Cycle.

1. The main burner flame is extinguished.
   a. The pilot gas remains lighted.
2. The warp switch cools.
   a. The main gas valve cannot open until about two minutes have elapsed after the power is restored.

Failure to Ignite Pilot Gas.

1. The same results as *No Gas on Start* occur.

Limit Switch Opens (in the thermostat circuit).

1. The main burner flame is extinguished.
   a. The main burner gas cannot flow.

Warp Switch Heater Burns Out.

1. Step 5 of the *Normal Lighting Sequence* does not occur. Subsequent steps, therefore, cannot occur.

See Fig. 10-4 for a schematic of the normal lighting sequence.

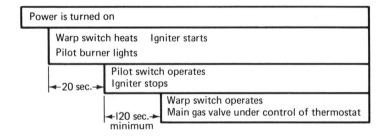

Fig. 10-4. Steps of normal lighting sequence.

The low voltage electricity is supplied by a Class II transformer and must supply a minimum of 25V, 60 cycles during the ignition and running cycles when energized by its rated primary voltage.

Repairs of this control cannot be made in the field. These controls, when requiring attention, should be returned to the factory and a new one should be installed according to the manufacturer's specifications.

A typical unit wiring diagram is shown in Fig. 10-5.

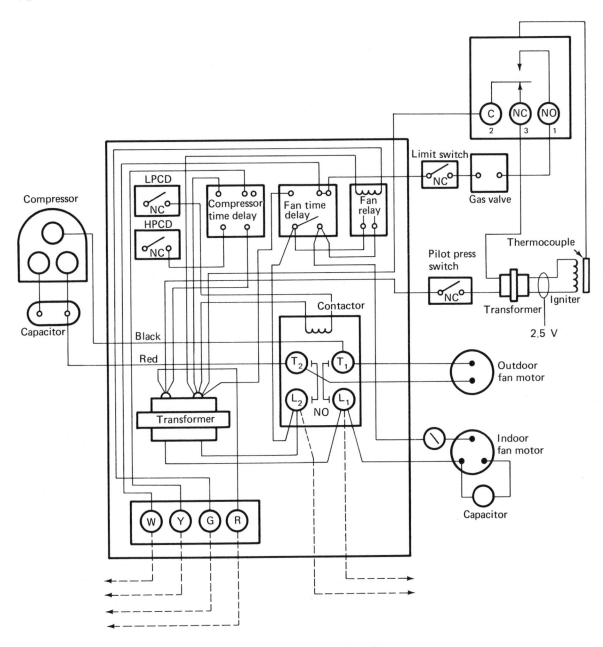

Fig. 10-5. Typical unit wiring diagram.

The spark pilot igniter automatically and instantly lights or relights the pilot gas. This is a solid state control and is new on the market and to the air conditioning trade. It is universal in application and should be given careful consideration when pilot troubles occur.

In operation the lighting or relighting of the pilot flame is accomplished by a spark across a gap of approximately 5/32 in. from the electrode tip to the grounded surface of the pilot burner. When the flame is established, the pilot flame conducts a current to the grounded pilot burner and a solid state switch in the unit turns off the spark. If the pilot flame is extinguished, the current to ground is interrupted and the solid state switch turns on the spark, which sparks at a rate of about 100 times per minute and thus relights the pilot well before the safety system drops out.

If it is desirable to shut down the pilot operation at any time it will be necessary to install a switch in the power line to the control board circuit.

The installation and wiring of this unit should be done according to the manufacturer's recommendations for satisfactory results.

These automatic ignition systems are used on heating equipment whose remote locations or environmental conditions make automatic lighting desirable. Typical applications include unit or duct heaters and outdoor heating units.

A *standing pilot* is a pilot which when lighted will remain so until the gas supply to it is stopped.

An *intermittent pilot* is one that is ignited each time the control, or thermostat, demands heating to be supplied to the space.

QUIZ 10

1. What are the main types of automatic pilot igniters?
2. What is the operating voltage of the igniters?
3. How wide is the gap on spark igniters?
4. Will the limit switch stop the igniters?
5. How long is the delay before the warp switch contacts reverse an interruption of gas?
6. How long will it be before a second attempt is made to restart the pilot when no gas is available on the start up?
7. On power interruption only the _____ gas flows.
8. Is it possible to have a gas supply to the main burner without ignition available?
9. Automatic pilot igniters have a _____ similar to a conventional type of pilot safety.
10. What type of transformer is used on automatic pilot igniters?

11

BOILER CONTROLS

Control of the energy output of a boiler is obtained by regulating the input. Small oil or gas fired boilers are regulated by operating controls that turn the burner on or off.

A *boiler control* may be defined as any control that provides safe, automatic, and economical operation of a boiler.

DEFINITION

One way to prevent damage to a boiler from overheating is to stop the burner. In Fig. 11-1 this is accomplished by a float-operated automatic low water fuel cut-off. This unit is designed to operate on boilers with pressures up to about 30 psi. It is mounted on the boiler so the switch will interrupt current to the main gas valve when the boiler water level drops to an unusually low point.

WATER LEVEL
CONTROL

Figure 11-2 shows a typical installation of a low water cut-off on a hot water boiler. There are several locations and piping arrangements available for this type of installation. Because there is no normal water line to be maintained in the boiler, any location of the control above the lowest permissible water level is satisfactory.

Thus a hot water boiler can operate in the usual manner and yet be safeguarded against the emergency condition of low water.

The construction of a hot water boiler is essentially the same as a steam boiler. Most of the conditions causing low water to occur in a steam boiler will also hold true for a hot water boiler.

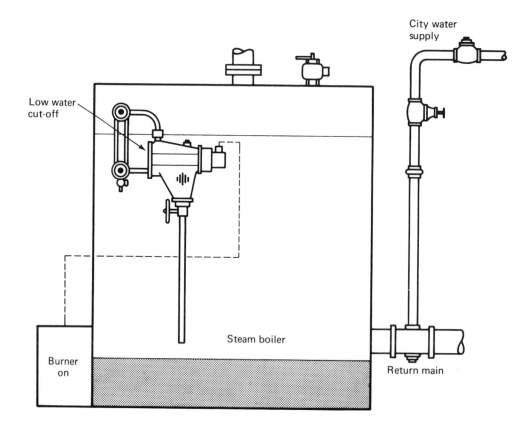

Fig. 11-1. Float operated low water cut-off.

The lowering of the boiler water line and the simultaneous lowering of the water line in the float chamber cause the float to drop, thus opening the electrical circuit and stopping the automatic burner (Fig. 11-3).

Here then is a basic safety control—a means of stopping the automatic firing device if the water in the boiler drops below the minimum safe level.

The following statement appears in a booklet, "Recommended Practices for Installation," published by a leading utility company. Their experience in the heating field prompts them to write:

> A low water cut-off which will cut off the fuel supply before the water level reaches a low danger point, or a water feeding device with cut-off, shall be attached to all steam and hot water boilers (*McDonnell Basic Safety Controls for Hot Water Space Heating Boilers,* Bulletin No. P-30C, Chicago, Ill., 1962, p. 6).

If we could rely absolutely on the low water cut-off to stop the automatic burner each time a low water condition developed, then the

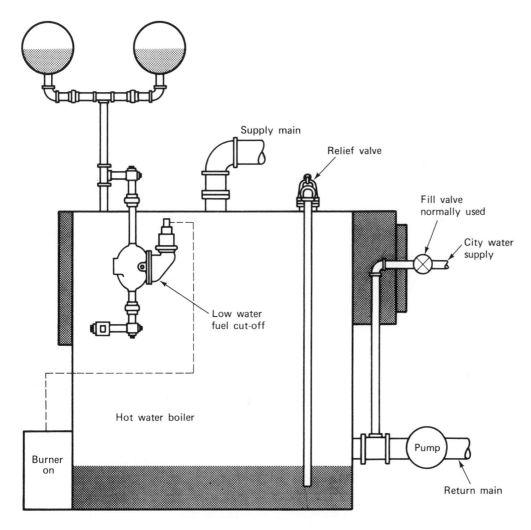

Fig. 11-2. Installation of low water cut-off.

problem would be solved completely. However, experience has proved that under certain circumstances the low water cut-off cannot fulfill its duties.

The final recommendation, which covers all installations and provides the most complete measure of safety, is to use a combination boiler water feeder and low water cut-off (Fig. 11-4).

This provides:

1. The mechanical operation of feeding water to the boiler as fast as it is discharged through the relief valve.
2. The electrical operation of stopping the burner when low water occurs.

This combination of mechanical and electrical safeguards is the best and most complete recommendation for boilers.

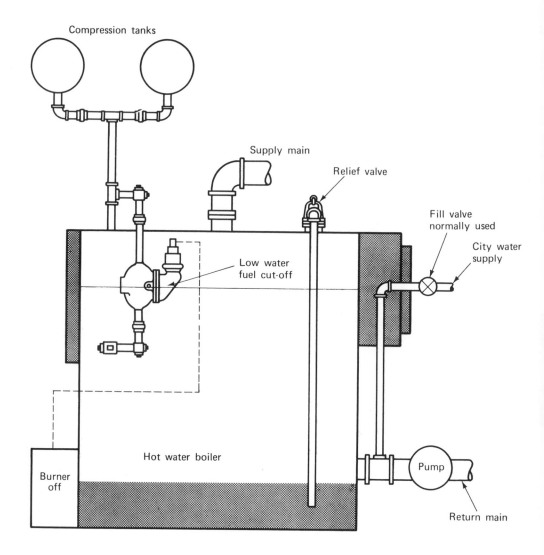

Fig. 11-3. Low water line stops burner

The diagram in Fig. 11-5 is typical of the operation of many manufacturers' electrical switches.

In Fig. 11-5(b) the NC contacts complete the control circuit and allow the burner to operate on demand from the temperature or pressure control.

In Fig. 11-5(c), the diagram shows that when a low water level condition occurs, the control circuit is opened and the N.O. alarm contacts are made, sending a danger warning to the operating engineer.

In normal operation the control of the burner is accomplished by either a temperature or a pressure control, depending on the type of system being controlled.

In the case of the temperature control the sensing element is installed in the boiler proper through the openings provided by the boiler manufacturer. The switching device may be either SPST snap

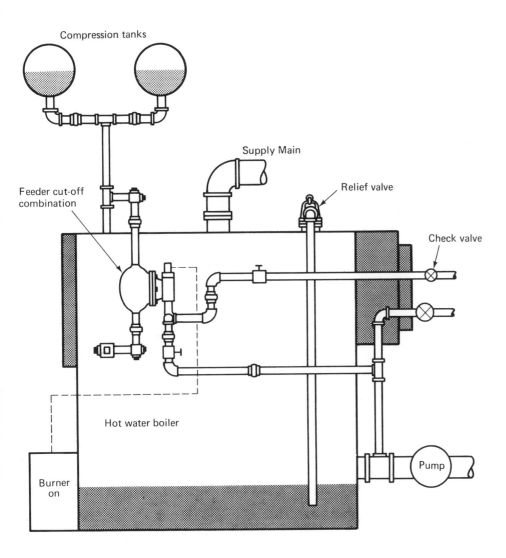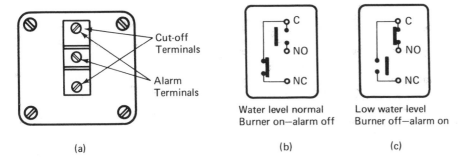

Compression tanks

Supply Main

Relief valve

Feeder cut-off combination

Check valve

Hot water boiler

Burner on

Pump

Fig. 11-4. Combination feeder and low water cut-off.

Cut-off Terminals

Alarm Terminals

(a)

C
NO
NC

Water level normal
Burner on—alarm off

(b)

C
NO
NC

Low water level
Burner off—alarm on

(c)

Fig. 11-5. (a) Switch terminal locations. (b) Water level normal. Burner on—alarm off. (c) Low water level. Burner off—alarm on.

action switch or a mercury tube. Both are actuated by a helically wound bimetal. The temperature control is called an *aquastat* while the pressure control is called simply a boiler pressure control. The aquastat is shown in Fig. 11-6.

Fig. 11-6. Aquastat (Courtesy of Honeywell, Inc.).

The boiler pressure control is used on steam boilers and is mounted above the boiler proper to sense the pressure at the most critical point of the system (Fig. 11-7).

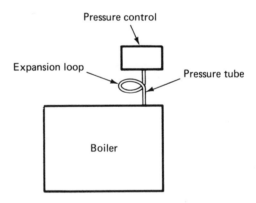

Fig. 11-7. Boiler pressure control installation.

Both controls are wired into the control circuit in the same manner; a wiring diagram is shown in Fig. 11-8. Also, both controls are assisted by a high limit control to stop burner operation if the pressure or temperature exceeds the maximum operating conditions set forth for a given boiler.

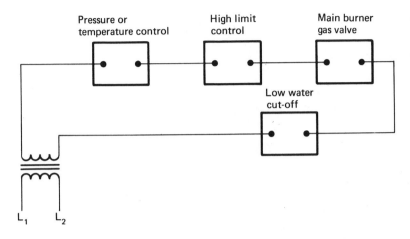

Fig. 11-8. Basic boiler wiring diagram.

The high-low-off type of combination diaphragm valve provides all manual and automatic control functions required for operation of gas fired heating equipment. These devices have an internally vented diaphragm-type main control valve and also a separate thermomagnetic safety valve with a pilot gas adjustment and a pilot filter. Most models include a pressure regulator function for use with natural gas (Fig. 11-9).

The main line gas valve diaphragm opens and closes in response to an aquastat or a boiler pressure control. The main line diaphragm valve functions simultaneously as a pressure regulator on natural gas models. The LP gas models are not equipped with a pressure regulator.

When the electrical circuit is completed between the C and W_1 terminals of the valve, the valve automatically opens to the preset low-fire position. Then, when the circuit is completed between C and W_2, the valve opens to the high-fire position. The low-fire to high-fire shift is accomplished by a heat motor in the valve operator. Time must be allowed for this heating action to be accomplished (Fig. 11-10).

This type of valve is used when the full BTU rating of the boiler is not needed but when some heating is required.

HIGH-LOW FIRE VALVES

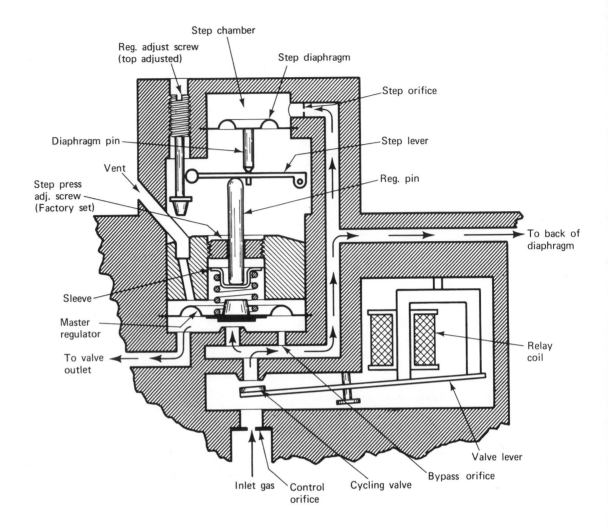

Fig. 11-9. Cut-away view of high-low fire valve.

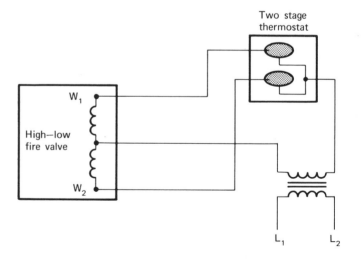

Fig. 11-10. High-low fire wiring diagram.

QUIZ 11

1. What is the purpose of a water level control?
2. Under what condition does the water level control react?
3. What is the most recommended boiler safety control?
4. Water level controls are designed to operate under pressures up to
   _____ .
5. What is the basic difference between hot water and steam boilers?
6. What two functions does a combination boiler water feeder and low water cut-off provide?
7. The combination boiler water feeder and low water cut-off can stop the burner and _____ simultaneously.
8. Normal burner operation is controlled by a _____ or _____ .
9. For what reason are high−low−off gas valves used on boilers?
10. Is a gas pressure regulator used on both LP gas and natural gas?

12

OIL BURNER CONTROLS

Of the three most used fuels (oil, gas, and electricity) oil is not, by far, the least used. There are approximately 65 million users of oil heating equipment. Because of the popularity of fuel oil it is necessary to include controls for safe, automatic operation in a book of this type. Fuel oil burners use a few controls that are not uncommon to gas equipment, such as the fan control, limit control, thermostat, and so on.

An *oil burner* is a device that supplies a mixture of atomized fuel oil and air, under pressure, to the combustion chamber, where the majority of combustion takes place.

DEFINITION

The *stack control* is simply a combustion thermostat used to sense changes in the temperature of the flue gases. These changes in flue gas temperature operate the stack contacts of the stack control; through the action of the thermal element they control the operation of the safety switch in the relay unit, thus assuring safe starting and proper burner operation. Both recycling and nonrecycling combinations of stack controls are available. These units are mounted directly on the flue vent with the sensing element inside the flue pipe. They should be mounted as near as possible to the furnace or burner and at least two feet from the draft regulator (Fig. 12-1).

STACK CONTROL

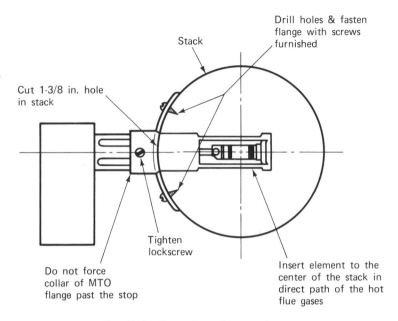

Stack

Drill holes & fasten flange with screws furnished

Cut 1-3/8 in. hole in stack

Tighten lockscrew

Do not force collar of MTO flange past the stop

Insert element to the center of the stack in direct path of the hot flue gases

Fig. 12-1. Typical stack control mounting.

OPERATION

To start burner (Fig. 12-2; *Caution*: be sure that the combustion chamber is free of oil):

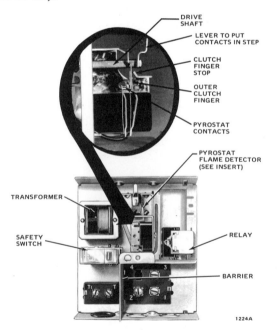

DRIVE SHAFT

LEVER TO PUT CONTACTS IN STEP

CLUTCH FINGER STOP

OUTER CLUTCH FINGER

PYROSTAT CONTACTS

PYROSTAT FLAME DETECTOR (SEE INSERT)

TRANSFORMER

SAFETY SWITCH

RELAY

BARRIER

1224A

Fig. 12-2. Components of stack control (Courtesy of Honeywell, Inc.).

1. To put contacts "in step," pull drive shaft level outward ¼ in. and release slowly.
2. Move red reset lever to right and release.
3. Open hand valve in oil supply line.
4. Set thermostat to call for heat.
5. Close electrical switch. Burner should start.

IGNITION TIMING

If the relay drops out too soon after burner starts, adjust ignition timing lever toward MAXIMUM.

SCAVENGER TIMING

1. With the burner on, the drive shaft carries the clutch finger outward. The stop arm halts the clutch finger but the drive shaft moves a small amount farther. This override is necessary for proper sequencing. *Note:* If the clutch finger does not reach the stop arm with the recycle lever at the minimum setting, the bimetal is not getting enough heat. The stack control must be relocated.
2. Allow the burner to run a few minutes, then open and close the electrical line switch. The burner should stop at once.
3. The burner should start in about one minute.
4. If the burner restarts too soon, proceed as follows:
   a. Open the line switch; wait five minutes for cooling.
   b. Move the recycle lever outward one notch. Close the line switch to start the burner, and repeat steps 1, 2, and 3.
   c. Repeat steps a and b until timing is satisfactory.

To check the stack switch out, use the following procedure to verify the safety features:

1. Flame failure:
   a. Test for recycling by shutting off the oil supply hand valve while the burner is operating normally. Restore the oil supply after the burner shuts off. After a short scavenger period, the stack control restarts the burner.
   b. Test for safety shutoff after flame failure by shutting off the oil supply hand valve while the burner is operating normally. When the burner shuts off do *not* restore the oil supply at this time. The stack control will attempt to restart the system

after a scavenger period; then in approximately 80 sec, the safety switch locks out. Reset the safety switch, and the burner will restart.

2. Ignition Failure:

a. Test by turning the oil supply off while the burner is off. Run through the starting procedure, omitting Step 3. A lockout will occur. Reset the red safety switch.

3. Power Failure:

a. Turn off the power supply while the burner is on. When the burner stops, restore the power and the burner will restart after a scavenger period.

A typical wiring diagram is shown in Fig. 12-3.

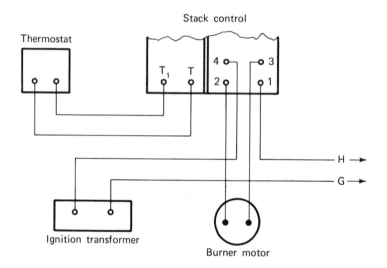

Fig. 12-3. Typical wiring diagram.

Another type of combustion thermostat is burner mounted. This thermostat is a hermetically sealed switch that opens and closes its contacts in response to temperature changes produced by the radiant heat of an oil burner flame. It is not a light sensitive device and is not affected by normal soot deposits. Its small size permits easy mounting on the blast tube of the burner, where it will quickly sense the radiant heat of the flame pattern (Fig. 12-4).

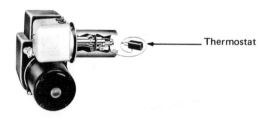

Fig. 12-4. Burner mounted combustion thermostat (Courtesy of ITT General Controls).

This control monitors the burner flame. In the starting position the contacts must be in the closed position. The control contacts will remain in the closed position until there is enough radiant heat from the flame to open its contacts.

When the proper flame is established, a rise in temperature of the sensing face will open the contacts of the control within seconds and de-energize a safety switch heater allowing normal start and run. In the event of loss of the flame during a burner cycle, the resultant drop in temperature will close the contacts and energize the safety circuit of the relay.

After a normal burner run the combustion control contacts will return to the closed or starting position. If an immediate start is called for by the thermostat, this results in an enforced purge period while the combustion contacts and ignition switch are being restored to the closed or starting position.

BURNER SAFETY CONTROLS

These controls are the same safety controls that are used on warm air heating systems and steam or hot water boilers. They will stop the burner in the case of abnormal pressures or temperatures resulting from insufficient water or air flow and are wired into the system in the same manner.

A typical wiring diagram is shown in Fig. 12-5.

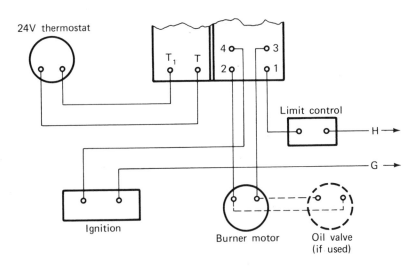

Fig. 12-5. Wiring diagram with burner safety control.

QUIZ 12

1. The stack control is essentially a _____ .

2. Where is the stack control mounted?

3. Name the two types of stack controls.

4. Are safety devices incorporated in the stack control proper?

5. Are some of the safety devices mounted and wired externally to the stack control?

6. Is it possible to check each function of the stack control without checking the whole control?

7. What is another name for the stack control?

8. One type of stack control is mounted on the _____ and the other is mounted in the _____.

9. Explain the ignition failure test procedure.

10. What should be done if the ignition timing relay drops out too soon?

13

MODUTROL MOTORS AND STEP CONTROLLERS

The on-off control is probably the simplest and least expensive type of control. When the thermostat senses a need for conditioned air, it turns on; when the need is satisfied, it turns off.

A significant improvement over the on-off control can be produced by dividing the conditioning load into a number of separate elements and keeping enough of them on continuously to provide an even flow of conditioned air to the area. This is done by using modutrol motors and step controllers to turn on the number of conditioning elements required to maintain a constant temperature.

DEFINITIONS

A *modutrol motor* is a motor that, when the proper signal is received, will change position in response to the signal.

A *step controller* consists of a series of switches (steps) operated by a motor.

OPERATION

The modulating motor circuit (Fig. 13-1) operates to position the controlled device (usually a damper or motorized valve) at any point between fully open and fully closed; this will proportion the delivery to the need as indicated by the controller mechanism.

The power unit is a low voltage capacitor motor which turns the motor drive shaft by means of the gear train. Limit switches are operated by the motor so that the rotation is limited to 160°. The gear train and all other moving parts are immersed in oil to eliminate the need for periodic lubrication and to insure long, quiet service.

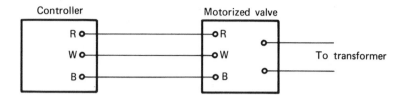

Fig. 13-1. Modulating motor circuit—field wiring.

The power unit is started, stopped, and reversed by the single pole-double throw contacts of the balancing relay (Fig. 13-2). The balancing relay consists of two solenoid coils with parallel axes, into which are inserted the legs of a U-shaped armature. The armature is pivoted at the center so that it can be tilted by the changing magnetic flux of the two coils.

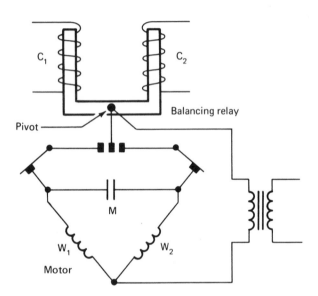

Fig. 13-2. Diagram of balancing relay and motor circuit.

A contact arm is fastened to the armature so that it will touch one or the other of the two stationary contacts as the armature moves back and forth on its pivot. When the relay is in balance, the contact arm floats between the two contacts, touching neither of them.

A balancing potentiometer is included in the motor. The potentiometer is electrically identical to the one in the thermostat. The finger is moved by the motor shaft so that it travels along a coil and establishes contact wherever it touches (Fig. 13-3).

Figure 13-2 illustrates how a balancing relay is made. As the relay is used in the modutrol circuit (Fig. 13-4), the amount of current passing through the coils is governed by the relative positions of the controller potentiometer and the motor balancing potentiometer. Thus, when equal amounts of current are flowing through both coils of the balancing relay, the contact blade is in the center of the space between

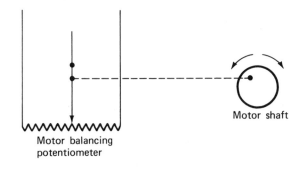

Fig. 13-3. Schematic diagram of balancing potentiometer.

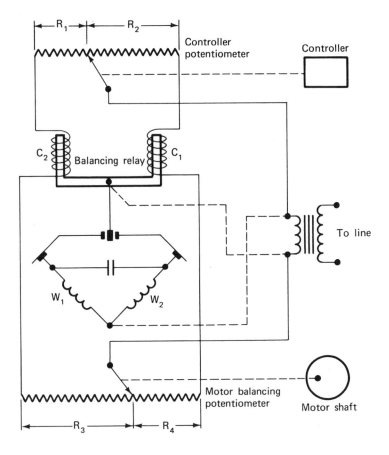

Fig. 13-4. Diagram of control circuit in balanced condition.

the two contacts and the motor is at rest. When the finger of the controller potentiometer is moved, more current flows through one coil than the other and the relay is unbalanced. The relay armature is then rotated so that the blade touches one of the contacts and the motor runs in the corresponding direction.

The contact made by the balancing relay can only be broken if the amount of current flowing through coil C_1 is made equal to that amount flowing through coil C_2. This is brought about by the motor balancing potentiometer linked to the motor shaft. As the motor rotates, it drives the finger of the motor balancing potentiometer toward a position which will equalize the resistances in both legs of the circuit.

There is a definite position of the finger for each position of the motor shaft throughout its 160° arc of travel. For example, when the motor shaft is 40° from one of its extremes (25 per cent of its arc) the finger is at 33¾ ohm, a value that lies 25 per cent of the distance from the corresponding extreme of the coil resistance.

Figure 13-4 shows an instantaneous condition in which the current is flowing from the transformer, through the potentiometer finger, and down through both legs of the circuit. In the positions shown, the thermostat potentiometer finger and the motor balancing potentiometer finger divide their respective coils so that $R_1 = R_4$ and $R_2 = R_3$. Therefore $R_1 + R_3 = R_2 + R_4$ and the resistances on both sides of the circuit are equal. The coils C_1 and C_2 of the balancing relay are equally energized and the armature of the balancing relay is balanced. The contact arc is floating between the two contacts, no current is going to the motor, and the motor is at rest.

Modutrol motors may be used for:

1. Air flow diversion - Where a parallel air flow pattern is used, a diverting damper is used to direct the air flow through either the heating or cooling unit.
2. Air flow changeover - Where a resistance damper is used to decrease the air flow on the heating cycle.
3. Ventilation control - Where provision is made for introducing outdoor air into the system during the cooling season but not during the heating season.
4. Zoning - Where both heating and cooling may be desired at the same time but in different areas.
5. Valve operation - Where steam or water may need to be directed in varying amounts such as: steam or hot water heating coils or condenser cooling water to a condensing unit.

Step control is appreciably better than on-off control because it meets load requirements continuously (Fig. 13-5). Temperature fluctuations would be substantially greater with on-off control under identical circumstances. However, the modulating effect of step control is strictly dependent on the number of steps provided. Ideally, a sufficient number of steps should be provided to hold the coil air temperature rise to within 5°F per step.

A step controller consists of a series of switches operated by an actuator. The actuator seeks a switch position corresponding to the controller requirements, thereby closing the correct number of

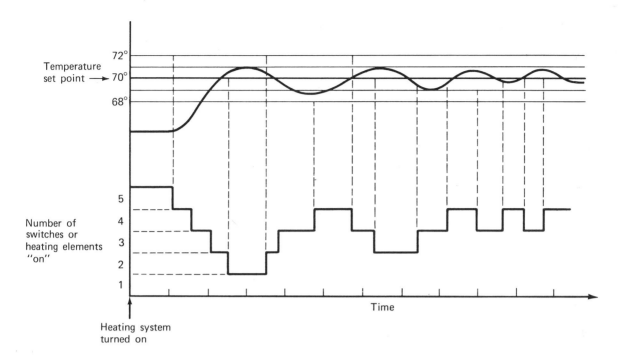

Fig. 13-5. Temperature variation with step control.

switches. The switches can directly energize the conditioning elements or control these elements through contactors.

The actuators on these units usually have a modulating type of motor and operate from a modulating thermostat. Some are of the proportioning type and are controlled by a staged (step) controller.

Step controllers may be used for control of any electric heating unit containing elements which can be divided into separate circuits. Also, they may be used to start or stop refrigeration compressors that are connected in tandem to produce the necessary refrigeration. The control of compressor unloaders is also accomplished with this control.

QUIZ 13

1. A modutrol motor will operate through _____ degrees.
2. What voltage is a modutrol motor?
3. Step controllers should hold a temperature rise of _____ degrees.
4. Which way will a modutrol motor operate if more current is directed through C_1?
5. The resistances are equal on both sides of the modutrol circuit when $R_1 =$ _____ and $R_2 =$ _____.
6. Current is admitted to the modutrol motor through the _____.
7. What are two uses of the modutrol motor?

8. The step controller may be used as a modulating control by the use of sufficient _____.

9. Can the step controller be used as either a pilot duty or line voltage control?

10. Step controllers are used on refrigeration compressors to _____ .

14

TIMERS
AND TIME CLOCKS

Timers and time clocks are devices used for the automatic and economical operation of air conditioning and refrigeration equipment. They are designed with a variety of contact arrangements and contact current ratings. If the time control you need is not immediately available, most manufacturers will welcome the opportunity to satisfy your time control requirements specially.

The *time clock* is an electrically operated control used to govern the various functions of equipment over a given period of time (Fig. 14-1).

Fig. 14-1. Time clock (Courtesy of International Register Co.).

The *timer* is similar to a time clock except that operations are timed regardless of the time of day. It measures the time elapsed before, during, or after an operation or the relationship between the on and off periods (Fig. 14-2).

Fig. 14-2. Timer (Courtesy of International Register Co.).

OPERATION   To help us understand how these controls operate we should first become acquainted with the interior components of a time clock.

The gears and parts are heavy and rugged enough to withstand prolonged normal usage (Fig. 14-3); they are designed and built to function properly even under adverse conditions.

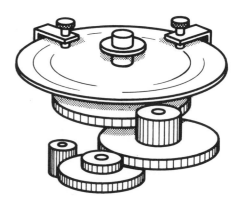

Fig. 14-3. Gears.

The high powered switching mechanism (Fig. 14-4) is made of channeled spring brass U-beam blades operated by a rugged cam to give instant and positive make and break action. The contacts are usually

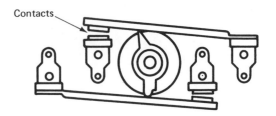

Fig. 14-4. Switching mechanism.

self-cleaning and made of a special alloy to prevent pitting. They are rated to carry an inrush current ten times their normal amperage rating without arcing or sticking.

Time clocks are usually powered by heavy duty industrial-type motors. (Fig. 14-5) These motors are a synchronous type which is extremely quiet and self-lubricating. They never need service or attention and are practically immune to adverse temperature and humidity conditions.

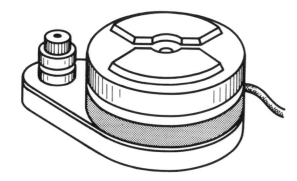

Fig. 14-5. Industrial motor.

The terminal board (Fig. 14-6) is designed for fast and easy wiring; there is plenty of room for hands and wiring.

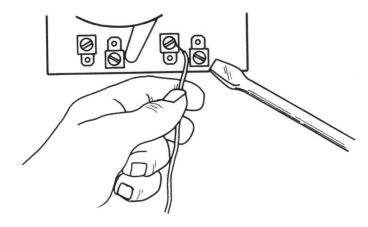

Fig. 14-6. Terminal board.

The control dial (Fig. 14-7) is divided accurately into units of time. Most are painted with contrasting colors to facilitate the placement of trippers. The trippers are adjustable to provide the desired on and off periods. These dials provide 24 hour coverage. The trippers actuate the contacts and are fastened to the rotating dial by thumbscrews. The dial is set at the correct time by pushing the dial toward the rear of the control and rotating until the correct time corresponds to the time indicator. It is necessary, however, to consult the manufacturer's recommendations before attempting to set the dial.

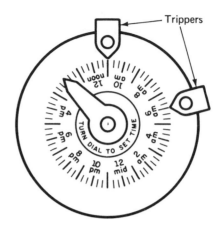

Fig. 14-7. Control dial.

The skipper dial (Fig. 14-8) allows operations to be skipped on selected days of the week, by inserting screws into the proper holes in the dial for the days on which the operation is to be omitted.

Fig. 14-8. Time clock showing skipper dial (Courtesy of International Register Co.).

The hand trip (Fig. 14-9) allows the operation of the on and off switch without disturbing the dial settings. The trip is moved to the "on" position to determine if the system is functioning properly.

In operation the time clock motor is wired in parallel with the switching contacts. As the motor rotates the dial and trippers, the contacts are opened or closed, depending on the operation desired at that time. A simple diagram is shown in Fig. 14-10. Most time clocks are provided with a spring wound carry-over mechanism. The carry-over keeps the time clock on schedule for 36 hours during power failures. When the power resumes, the carry-over automatically rewinds itself,

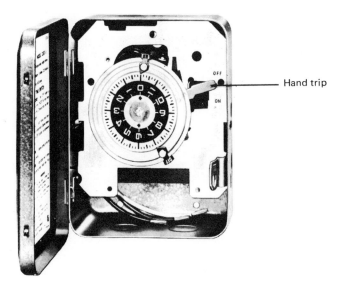

Fig. 14-9. Hand trip (Courtesy of International Register Co.).

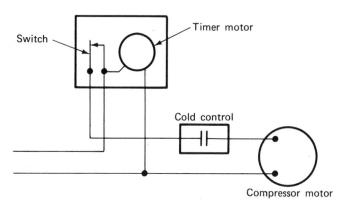

Fig. 14-10. Simple wiring diagram of defrost circuit.

thus providing an ideal control in areas where power breakdown frequently occurs.

A time clock provided with all the devices mentioned previously can be used to control almost any operation desired. The following are some of the more common uses of time clocks:

1. Heating control
2. Ventilation control
3. Air conditioning control
4. Defrost control

Timers provide automatic control over a great variety of functions. Through the use of relays they may be used to provide the same functions as the basic time clock. The 24 hour dial (Fig. 14-11) is divided into half-hour increments with a permanently attached "on" and "off" tripper permitting one on-off control.

Some suggested uses for the timer are:

Fig. 14-11. Timer dial (Courtesy of International Register Co.).

1. Time delay relays
2. Motor control (to 1 HP)
3. Pump control
4. Heating control
5. Domestic refrigeration control

QUIZ 14

1. List the main parts of the time clock.

2. List the auxiliary parts of the time clock.

3. When would the spring carry-over operate in a time clock?

4. What part of the time clock actuates the contacts?

5. What does the hand trip do?

6. How is the time clock motor wired in relation to the contacts?

7. List several uses for a time clock.

8. How does a timer operate?

9. What type of motor is used in the time clock?

10. What control is needed to allow the timer to be used as a basic time clock?

15

DEFROST SYSTEMS

When a refrigeration unit producing below freezing temperatures is operated for a period of time, a layer of frost or ice is built up on the evaporator. This layer of ice acts as an insulator and decreases the efficiency of the equipment. To employ a person to keep watch and defrost the coils would be too expensive; thus automatic defrost systems are used.

Defrost systems are a variety of controls used to accomplish the automatic defrosting of refrigeration evaporators.

DEFINITION

The simplest defrost system uses the low pressure control to accomplish this goal. The low pressure control is set low enough to maintain the lowest temperature required in the refrigerated space. The low pressure switch has primary control over the compressor. Operation of the compressor is started only after the pressure and temperature of the evaporator have risen sufficiently to cause some of the ice to melt during each off cycle of the equipment.

OPERATION

This defrost system is not very efficient because not all of the ice is melted during each cycle. When the low pressure control is used the complete system must be shut down periodically to defrost the system manually. This system is not the most economical; it is also inconvenient.

The most economical and efficient system is one that completely defrosts the evaporator during each defrost cycle. These systems are usually time initiated and temperature terminated. Thus they allow a more even temperature during the busy time of the day, and the defrost period is accomplished during a period of less activity.

There are two main classifications of defrost systems: 1) the electric defrost and 2) hot gas defrost. The hot gas defrost system will be discussed first.

The hot gas defrost system is the most rapid and economical method of automatic, positive defrosting. Refrigerant, superheated by compression, is circulated through the evaporator to provide a continuous supply of defrosting heat.

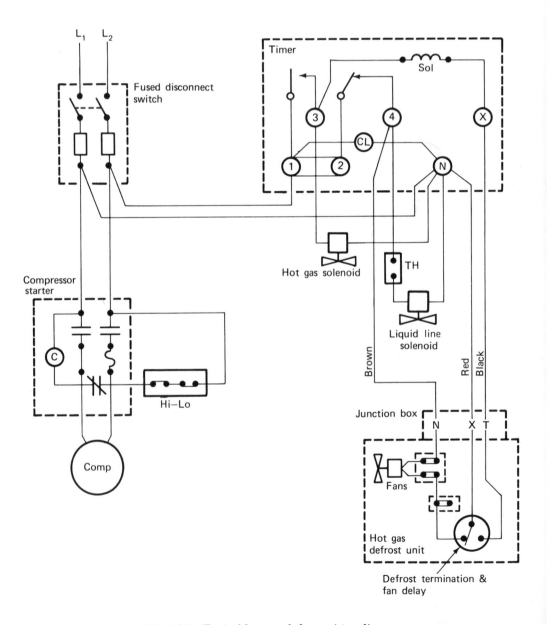

Fig. 15-1. Typical hot gas defrost wiring diagram.

Defrost cycles are time initiated and temperature terminated. The defrost intervals should be determined by the box usage so that defrosting is completed and the evaporator is free of frost just prior to periods of heavy usage. See Fig. 15-1 for a typical wiring diagram.

During the defrost cycle, the evaporator and condenser fans are turned off. A solenoid valve in the hot gas line opens, and the hot gas is circulated through the evaporator until it is completely defrosted.

At the proper temperature, the preset terminating thermostat ends the defrost period and resets the timer to the refrigeration cycle. Additional contacts are provided in the terminating thermostat for use as a fan delay. This prevents air circulation across the warm evaporator and avoids possible damage from rapidly expanding air in a closed box.

See Figs. 15-2 and 15-3 for more applications of defrost systems.

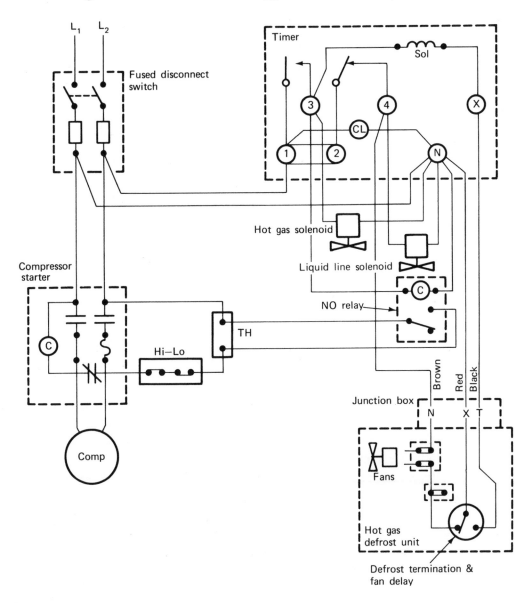

Fig. 15-2. Typical single unit wiring diagram.

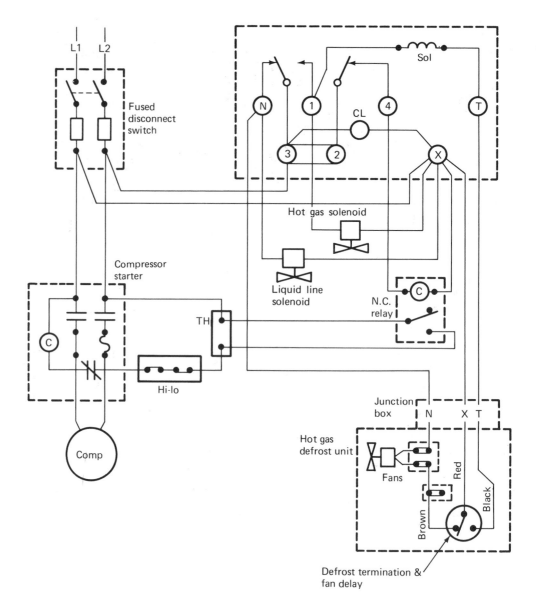

Fig. 15-3. Wiring diagram with positive compressor start at initiation of defrost.

When the electric defrost system is used, the compressor is shut down during the defrost cycle. An electric heater is energized to heat the evaporator and melt the ice. At the same time the evaporator and condenser fans are stopped. As the temperature is raised to a predetermined point, the defrost cycle is terminated and refrigeration resumes in the usual manner. See Fig. 15-4 for wiring diagrams.

A good quality time control should be used to initiate the defrost cycle in both the hot gas and the electric defrost systems.

The drain line should be heated in the refrigerated space to prevent freezing during the defrost cycle. This may be done with either an electric tape around the pipe or with the liquid line in contact with the drain line. The drain should be trapped outside the refrigerated space.

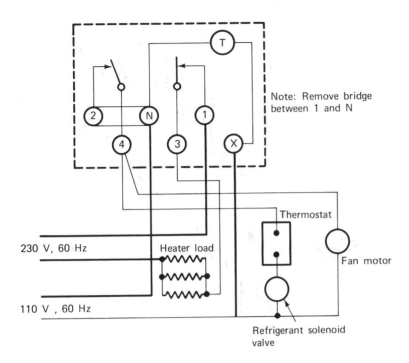

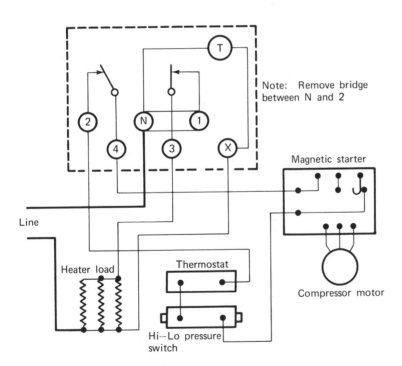

Fig. 15-4. *Typical wiring diagrams for electric heat defrost system.*

The liquid line solenoid valve should be of good enough quality to prevent liquid refrigerant from entering the low side during defrost regardless of the thermal expansion valve setting.

Typical examples for uses for defrost systems are:

1. Frozen food cabinets
2. Dairy cases
3. Vegetable cases
4. Beverage coolers
5. Ice cream display cases
6. Walk-in coolers

QUIZ 15

1. Name two types of defrost systems.

2. How are most defrost cycles initiated?

3. What is the simplest type of defrost system?

4. What precaution should be taken with the drain line?

5. What fan or fans are turned off during defrosting?

6. What control is used to terminate the defrost cycle?

7. When should the defrost cycle operate?

8. What objections are there to using the low pressure control for a defrost control?

9. How long should a defrost period be?

10. What type of solenoid valve should be used in the hot gas defrost system?

16

AIR FILTERING
CONTROLS

Atmospheric dust is a complex mixture of smokes, dusts, mists, and fumes, which are generally referred to as aerosols. A sample of atmospheric dust gathered at any given point will generally contain minute particles of materials from that locality together with other components that may have been dispersed by wind or air currents quite far from their original locations.

It is these particles which must be removed from the air in order for us to have clean, healthful structures in which to live and work. Proper control of these filtering systems is necessary to obtain healthful, efficient operation of the complete system.

Air filtering controls are the devices that oversee the equipment that actually filters the air in accordance with the demands of the space.

DEFINITION

One type of filtering control is a pressure sensing device which drops a red "clogged filter" flag into the window when a decrease of suction pressure, caused by filter obstruction, occurs (Fig. 16-1). It also actuates an indicator light on the system panel or special thermostat subbase. This indicates the need to clean or replace the air filters in the forced air heating, air conditioning, or heat pump equipment. A normally open switch completes a low voltage circuit to the indicator light (Fig. 16-2).

OPERATION

Fig. 16-1. Filter-flag indicator (Courtesy of Honeywell, Inc.).

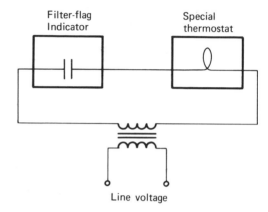

Fig. 16-2. Typical wiring diagram of filter-flag indicator.

This control is mounted on the downstream side of the furnace filter, directly on the fan compartment or at a remote location, by use of a piece of ¼ in. tubing (Fig. 16-3). The suction pressure at which the flag drops can be adjusted from a .1 in. to a .7 in. water column to compensate for different filter densities.

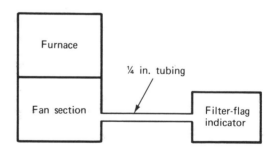

Fig. 16-3. Typical remote filter-flag indicator installation.

A more sophisticated filtering system is the electronic air cleaner. The electronic filter takes the place of the regular filter in an air conditioning system. It operates only when the fan in the heating or cooling system is operating. The same air that is heated or cooled is now electronically cleaned before it reaches the heating or cooling unit. This reduces the amount of dirt reaching the unit and thereby increases

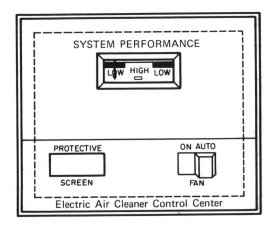

Fig. 16-4. Electronic filter control panel.

the efficiency of the entire system. This system is controlled by a control panel mounted on a wall near the thermostat (Fig. 16-4).

The following is a brief description of how airborne dirt is electronically removed (Fig. 16-5).

1. Airborne particles are carried to the electronic air cleaner along with the air circulated by the heating or cooling system.
2. Lint, feathers, animal hair, and other large particles are caught by the protective screen.
3. Most of the airborne particles are too small to be stopped by this screen and pass into the ionizing section of the electronic cell where they are given an intense electrical charge.
4. As the air carries these charged particles into the collecting section of the electronic cell, they are hurled against metal plates by the force of a powerful electrical field. This process is similar to the way a magnet attracts iron filings.

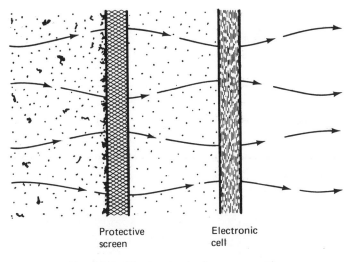

Fig. 16-5. Electronic air cleaner operation.

5. These particles cling to the metal plates so that only clean air enters the fan compartment of the heating or cooling system before recirculated through the house.

6. As the cleaned air circulates through the house and back to the electronic air cleaner, additional new airborne particles are constantly being carried to the air cleaner to be eliminated. These new particles are continually entering the house through doors and windows. New particles are also being generated inside the house, and existing dirt is stirred up by activity.

It is important to remember that the more the air cleaner is operated the cleaner the air in the home will be. Many furnaces and cooling systems permit you to select constant air circulation through your entire house and air cleaner. Operating the fan full time is recommended to give the cleanest air possible.

The performance meter on the control panel indicates at a glance how effectively the electronic cell is operating.

1. Needle steady and at the center of the scale indicates normal operation at maximum efficiency (Fig. 16-6).

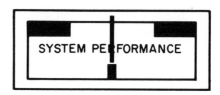

Fig. 16-6. Needle indicating operation at maximum efficiency.

2. If the needle moves out of the center position toward the right side, the needle is out of calibration (Fig. 16-7).

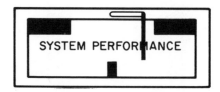

Fig. 16-7. Needle indicating meter out of calibration.

3. If the needle is at the extreme left position, the system is shut off (Fig. 16-8).

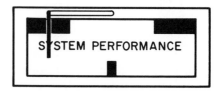

Fig. 16-8. Needle indicating system turned off.

4. When the needle slowly moves out of the center toward the left, the electronic cell is dirty (Fig. 16-9).

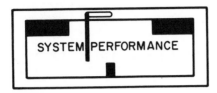

Fig. 16-9. Needle indicating dirty electronic cell.

5. A fluctuating needle indicates an arcing condition when accompanied by a snapping, crackling sound. If it is continuous, it indicates the electronic cell is dirty or has been damaged (Fig. 16-10).

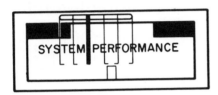

Fig. 16-10. Needle indicating dirty or damaged cell.

METER CALIBRATION

1. Turn on the air cleaner and fan (the needle moves toward the center of the meter scale when the air cleaner is operating).
2. When the needle is steady, turn the calibration screw (either clockwise or counterclockwise) until the needle is exactly over the bar at the center of the scale (Fig. 16-11).

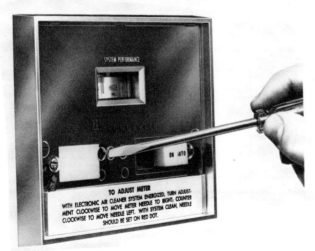

Fig. 16-11. Method of meter calibration (Courtesy of Honeywell, Inc.).

When the electronic filter is operating the electrical voltage is extremely high. It may be as high as 7,500 V, and therefore extreme caution must be exercised when working on these filter cells.

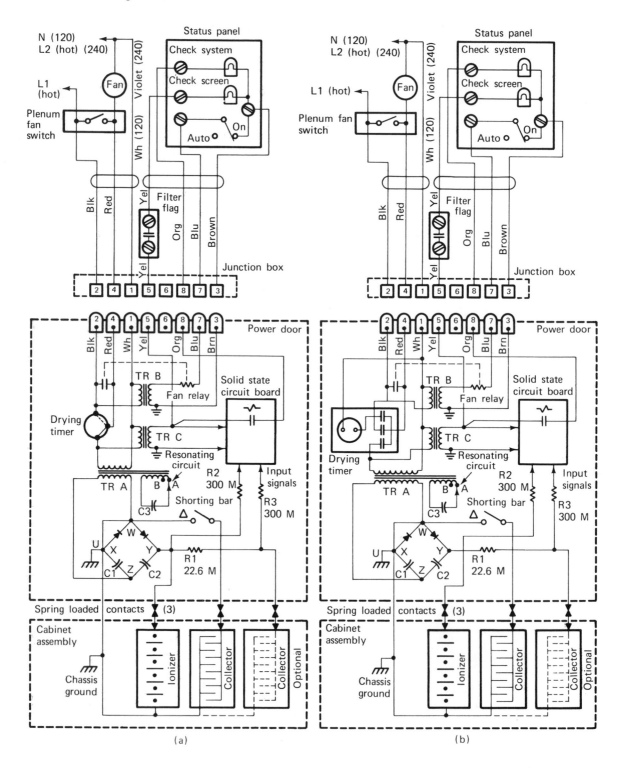

Fig. 16-12. (a) Wiring diagram of electronic filter with mechanical timer. (b) Wiring diagram of electronic filter with electrical timer.

The wiring on these units is very complicated and the diagrams shown above are only representative. The equipment manufacturer's recommendation must be followed when installing and servicing them (Fig. 16-12).

These filters are very beneficial to a person who is allergic to dust and pollen. Often these units are the only relief for hay fever and emphysema victims.

QUIZ 16

1. What is the simplest type of air filter control?

2. How is dirt removed by the electronic filter?

3. How high may the voltage be in the cells of electronic filters?

4. Outline the meter calibration procedure.

5. How does the clogged filter indicator operate?

6. Name two ways that a clogged filter may be detected with electronic filter controls.

7. What is the purpose of the protective screen?

8. On the control panel, what is indicated when the needle moves to the right?

9. What is indicated when the needle moves slowly to the left?

10. Can the fan be operated continuously from the control center?

17

DICTIONARY
OF TERMS

Ambient compensated Such a control is designed so that varying air temperature at the control does not materially affect control setting.

Ambient temperature Temperature of the surronding air in the immediate vicinity of the control body and power element system.

Adjustable differential A means of changing the difference between the control cut-in and cut-out settings.

Automatic changeover The change from one function to another, automatically. Otherwise, such changes would be accomplished by manual means.

Automatic recycle Control contacts return to original position automatically after actuation when the pressure or temperature returns to normal condition.

Auxiliary contacts A set of contacts to perform a secondary function usually in relation to the operation of the main contacts. For example, to cause an alarm to operate if the main contacts actuate.

Bellows A corrugated metallic diaphragm with a metal cup. Often a complete power unit is referred to as a bellows.

Break point Temperature at which all of the refrigerant charge of the element has completely vaporized.

At temperatures above the "break point" of the fill, there is a very little internal pressure change in the element—approximately 1 psi for each 10° F temperature change.

British thermal unit   Heat required to produce a temperature rise of 1°F in 1 lb. of water.

Contact rating   The capacity of an electrical contact to handle current and voltage.

Cross ambient fill   A vapor pressure element sufficiently large to assure liquid in the bulb regardless of whether the bulb is colder or warmer than the control ambient temperature. This occurs when temperatures at the control and capillary may be alternately above and below the desired temperature at the sensing bulb.

Current relay   A switching relay that operates on a predetermined amount of electrical flow (or lack of current flow).

Cut-in setting   The point at which the control electrical contacts close to "make" a circuit.

Cut-out setting   The point at which the control electrical contacts open to "break" a circuit.

Differential   The difference between the cut-in and cut-out settings of a control.

Differential screw   An adjusting screw used to change the difference between the cut-in and cut-out settings of a control.

Double pole   Two single pole contacts operating simultaneously.

Double throw   Contacts make in one direction but break simultaneously in the other direction.

Drop out voltage   The voltage or point at which the pull of the electromagnet is not strong enough to keep the armature seated.

Dry bulb temperature   The temperature indicated on a dry bulb thermometer. Indicates heat of air and water vapor mixture.

Dummy terminals   Extra terminals that do not connect to an electrical function on a control. Sometimes provided for wiring convenience.

Fail-safe control   A control design such that a component failure will cause the control to assume the safest action, protecting the system on which it is installed, usually a contact open condition.

Fixed differential   The difference between the control cut-in and cut-out setting is factory set and cannot be changed.

Fixed setting   Provides no convenient means for changing control settings after the control leaves the factory.

Flux   The electric or magnetic lines of force in a region.

Full load amperes   The amount of current in amperes in an electrical circuit when the load is operating in a full capacity condition.

Ground   Intentional or accidental connection from a power source to the earth, or a connecting body which serves in place of the earth, to complete an electrical circuit. Earth is considered zero potential.

Heat pump   An electrically operated device designed to extract heat from one location and transfer this heat to another location. A heat pump is used for both heating and cooling and reverses the refrigeration cycle when heating is needed.

Hermetically sealed   A compressor or other device that is enclosed in a gas-tight housing.

Horsepower rating   A rating in terms of a motor. Underwriters' Laboratories consider one horsepower equivalent to 746 watts. Most ratings now refer to electrical ratings in amperes.

Inductive rating   Maximum amount of amperes in a circuit when a conductor is in an electromagnetic field.

Inherent motor protection   A safety limit device built inside a motor or equipment. It protects for over-temperature, over-current or both.

Locked rotor amps (LRA)   Current which is required at the instant power is supplied to start a motor.

Manual reset   A control reset mechanism which requires a manual operation if the control cycle locks out on safety.

Modulating   A modulating control is one which corrective action is in small increments as opposed to complete on-off action.

Normally closed   This refers to a switch or a valve which remains closed when the device is not connected to a power supply or is de-energized.

On−off Used to describe control operations. Control is either on or off (two positions) as opposed to proportional or modulating.

Open circuit An electrical circuit which does not have a continuous path for current to flow. This open circuit may be caused by an open switch or a broken circuit, such as a blown fuse.

Overload protector A device which opens ungrounded conductors for protection against motor over current. It prevents unsafe running conditions and protects the motor from burn out.

Pilot duty rating An electrical rating applied to devices used to energize and de-energize pilot circuits such as the pull coil of a motor contactor.

Power The voltage used for actuating a device. A common use of power refers to electrical power or voltage.

Pull in voltage The voltage value which causes the relay armature to seat on the pole face.

Quick connects Terminals of a switch that are usually connected by "push" motion rather than the normal screw terminals.

Range The pressure or temperature operating limits of a control.

Range adjusting screw An adjusting screw used to change the operating set points of a control. Changes are limited to those within the control range.

Relative humidity Water vapor contained in a body of air as a percentage of the maximum water vapor density possible, which is 100 per cent relative humidity.

Set point Setting at which the desired control action occurs.

Single pole One set of two electrical contacts. These two contacts "make" and "break" on switch action.

Single throw Contacts "make" or "break" in only one direction of operation.

Terminal An electrical connection, such as a "screw terminal."

Terminate To complete an event or stop an operation.

Thermal relay A relay which is actuated by the heating effect of an electrical current. Sometimes referred to as a "warp switch."

Time delay device Designed to provide a time interval between operations of a device.

Underwriters' Laboratories A testing agency whose primary function is to assure that products are manufactured to meet specific safety standards. A listing of a product by Underwriters' Laboratories (UL) indicates the product was tested and the product met recognized safety requirements.

Warp switch A thermal relay switch is actuated by the heating effect of an electric current.

Wet bulb temperature The temperature indicated on a wet bulb thermometer. It indicates the total amount of heat in a mixture of air and water vapor.

18

ELECTRICAL SYMBOLS

An electric circuit can be complicated and sometimes difficult to understand. Engineers and technicians have adopted a common set of signs and symbols that are generally used in the drawing of circuits. A drawing using these symbols is called a *schematic*. Equipment manufacturers supply these schematics to aid the technician in locating troubles and in making the necessary repairs.

Schematics show the kinds of parts used and where they are connected in the circuit. The symbols sometimes have a letter identification by which they may be located in the parts list and their type and value found.

As the part is identified on the schematic, you will see a letter close to it, such as R, C, or L. This letter may also have a subscript. A subscript is a smaller letter located at the lower right side of the letter. It is used to distinguish one part from several others of the same kind.

The parts list is an important part of the description of an electrical circuit. An individual part can be identified by stating its value, its type, and its voltage rating, and by the manufacturer's stock number.

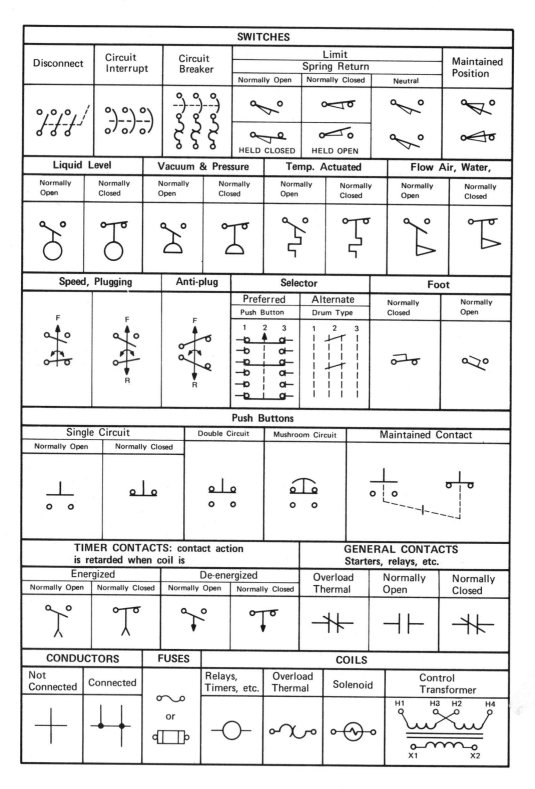

Fig. 18-1. Graphic electrical symbols.

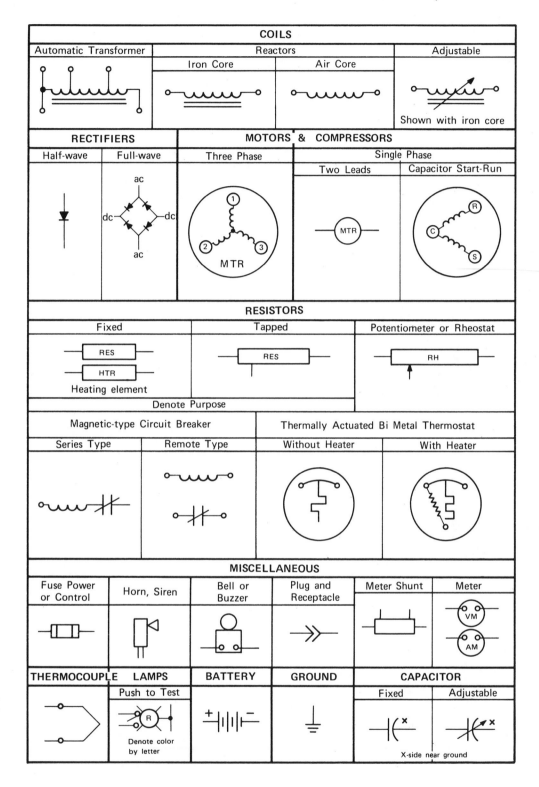

Fig. 18-1 (Cont.). Electrical diagrams.

19

SERVICE HINTS
ON CONTROLS

1. A control will cut out if:
   a. The cut-out temperature setting is below the lower limit of the refrigerating system.
   b. Control contacts are short circuited by faulty wiring.
   c. Too much ice is on the chilling unit.
2. A control will not cut in if:
   a. Power element has lost its charge.
   b. Open circuit exists in control wiring.

PILOT BURNER
SERVICE SUGGESTIONS

TROUBLE	CAUSE	REMEDY
	Pilot gas supply turned off	Turn on and relight pilot
	Pilot line contains air	Purge line
	Pilot burner orifice clogged	Clean and relight
Pilot cannot be lighted	Lighting knob not being depressed	Depress knob and light
	Lighting knob not set at pilot position	Set knob and relight
	Pilot gas flow adjustment closed off	Readjust and light pilot
	Lighting knob released too soon	Hold knob in position longer
	Reset button released too soon	Hold reset in position longer
Pilot goes out when reset knob is released	Thermocouple or powerpile is bad	Replace
	Pilotstat power unit bad	Replace pilotstat
	Bad connection on power unit	Clean and tighten connection
	Pilot flame improper size	Adjust flame
	Powerpile terminals shorted or loose	Clean and tighten connections or replace
	Pilot filter clogged	Replace or clean filter
	Pilot burner orifice clogged	Remove and clean orifice
	Gas supply pressure too low	Check for restriction or adjust main gas pressure regulator
	Pilot unshielded from excessive draft	Shield pilot
	Thermocouple or powerpile bad	Replace
	Pilotstat power unit bad	Replace pilotstat
Pilot goes out when system is in use	Power unit connection dirty, loose, or wet	Clean, dry, and tighten properly
	Pilot flame improper size	Adjust flame
	Powerpile terminals shorted or loose	Repair terminals
	Thermocouple cold junction too hot	Reposition thermocouple
	Pilot burner lint screen clogged	Clean
	Pilot unshielded from burner concussion	Shield or adjust main burner to stop concussion
	Pilot filter clogged	Clean or replace filter

PILOT BURNER
SERVICE SUGGESTIONS
(cont.)

TROUBLE	CAUSE	REMEDY
	Gas supply too low	Increase gas pressure
	Pilot blowing away from thermocouple	Shield pilot from draft
	Bad thermocouple or powerpile	Replace thermocouple or powerpile
	Bad coil in pilotstat	Replace coil or pilotstat
Pilot burning but unit turned off by safety control	Power unit connection bad	Correct connection
	Pilot flame improper size	Adjust pilot flame or clean pilot burner
	Powerpile terminals shorted or loose	Correct condition of connections
	Improper venting	Correct vent problem
	Too small gas line, or line restricted	Increase line size or clear restriction
	Pilot burner lint screen clogged	Clean screen
	Pilot burner orifice to large	Replace orifice with proper size
Pilot flame wavering and yellow	Excessive ambient temperature	Change location of pilot burner or provide more air
	Excessive ambient temperature	Change location of pilot burner
	Improper venting	Correct venting problem
	Gas supply pressure too low	Increase gas pressure
	Pilot unshielded from combustion products	Relocate pilot burner or shield
Pilot flame small, blue, waving	Pilot burner orifice clogged	Remove and clean orifice
	Pilot gas flow adjustment is closed off	Open adjustment
	Pilot burner orifice too small	Replace orifice with correct size
	Pilot filter clogged	Replace or clean filter
Pilot flame noisy, lifting, blowing	Pilot gas pressure too high	Reduce gas pressure
	Pilot burner orifice too small	Replace orifice with correct size
Pilot flame hard, sharp	Typical of manufactured butane-air and propane-air mixture	No correction
Pilot turn-down test bad	Pilot burner improperly located	Relocate

THERMOSTAT
SERVICE SUGGESTIONS

TROUBLE	CAUSE	REMEDY
Thermostat jumpered; system still doesn't work	Thermostat not faulty	Check elsewhere
	Limit control set too low	Raise setting
	Low voltage control circuit open	Find open circuit and repair
	Low voltage transformer bad	Replace transformer
	Main gas valve bad	Replace valve
	Bad terminals	Repair terminals
System works when thermostat is jumpered	Dirty thermostat contacts	Clean contacts
	Damaged thermostat	Replace thermostat
Room temperature overshoots thermostat setting	Thermostat located on a cold wall	Change location
	Thermostat wiring hole not plugged	Plug hole with putty
	Thermostat exposed to cold draft	Relocate thermostat
	Thermostat not exposed to circulating air	Relocate thermostat
	Mercury type thermostat not mounted level	Level thermostat
	Thermostat not in calibration	Calibrate, if possible. If not, replace
	Anticipator set too high	Reset anticipator
	Heating plant too large	Reduce BTU input
	No anticipator in thermostat	Replace thermostat
Room temperature doesn't reach thermostat setting	Thermostat not mounted level	Level thermostat
	Thermostat not calibrated properly	Recalibrate or replace thermostat
	Heating plant too small or underfired	Increase BTU input
	Limit control set abnormally low	Set limit control
	Thermostat exposed to direct rays of the sun	Relocate thermostat
	Thermostat affected by fireplace or heat from appliances	Relocate thermostat
	Thermostat located on warm wall, or near a register	Relocate thermostat
	Dirty thermostat contacts	Clean contacts
	Bad wiring on terminals	Repair terminals
	Dirty air filter	Clean or replace filter
Thermostat cycles unit too often	Heat anticipator set too low	Reset to correct amperage

THERMOSTAT
SERVICE SUGGESTIONS
(cont.)

TROUBLE	CAUSE	REMEDY
Thermostat doesn't cycle unit often enough	Thermostat not exposed to return air	Relocate thermostat
	Too small heating plant or plant is underfired	Increase BTU input
	Heat anticipator set too high	Set to correct amperage
	No anticipator on thermostat	Replace thermostat
	Dirty thermostat contacts	Clean contacts
Too much variation in room temperature	Thermostat not exposed to return air	Relocate thermostat
	Heat anticipator set high	Set to correct amperage
	Heating unit too large or overfired	Reduce BTU input
	No heat anticipator	Replace thermostat

MOTOR SERVICE
SUGGESTIONS

TROUBLE	CAUSE	REMEDY
Fan motor will not run	Bad motor bearings, starting switch, or burnt winding	Repair or replace motor
	Fan control contacts not completing circuit	Replace fan control
	Fan relay contacts not completing circuit	Replace fan relay
	Blown fuse	Replace fuse

SOLENOID VALVE
SERVICE SUGGESTIONS

TROUBLE	CAUSE	REMEDY
Valve will not open	Thermostat or other controller inoperative	Repair or replace inoperative controls
	Clocks, limit controls, or other devices holding circuit open	Check circuit for limit control operation, blown fuses, short circuit, loose wiring, etc.
	Solenoid coil shorted or wrong voltage	Replace coil
Valve will not close	Manual opening device holding valve open	Release manual opening device
	Valve not mounted vertically	Mount valve in horizontal line with solenoid in vertical position above pipe
	Bent or nicked plunger tube restricting valve opening	Replace plunger tube or solenoid
	Foreign matter in valve interior	Disassemble valve and clean thoroughly
	Limit controls in grounded side of circuit	Rewire control into hot side of circuit

DEFROST SYSTEMS
SERVICE SUGGESTIONS
(Electric heat or hot gas)

TROUBLE	CAUSE	REMEDY
Defrosts at wrong time	Defrost control set on wrong time	Set knob to indicate proper time of day
	Control may defrost at wrong knob position	Replace the control
	Timer motor current different than power supply	Replace control to correspond to power supply
Does not keep time	Clock does not run	Check for voltage at motor terminals; if power is found replace control. If not, check elsewhere
	Intermittent power supply or wrong wiring connections	Check power supply or correct connections
Will not defrost	Clock does not run	Check for voltage at motor terminals; if voltage is found replace control. If not, check elsewhere
	Clock runs but does not turn dial shaft	Replace control
	Defrost control bulb too warm	Properly locate and clamp control bulb
	Skipped a defrost cycle	Replace control if trouble persists
	Leak in fail-safe bellows	Replace control
Incomplete defrost	Excessive frost build-up between defrost periods	Check defrost operation by turning knob to just before defrost. Determine and correct excessive build-up. (Door may be loose or left open too much, uncovered foods, etc.)
	Defrost termination switch set too low	Raise temperature setting, (each 1,000 ft altitude lowers setting about 2.5°F)
	Partial lost fill of fail-safe bellows	If recalibration does not correct, replace control
	Bulb not properly attached to evaporator	Clamp bulb in clean area
	Cross ambient condition	Control must be warmer than bulb or tubing during operation
	Compressor cycles on overload	Replace starting relay or overload
	Defective defrost switch	Check defrost control contacts; if open replace control
	Cycling control lost fill	Check cycling control; if contacts are open replace control

TROUBLE	CAUSE	REMEDY
Defrost will not terminate	Cross ambient condition	Be sure control is warmer than the bulb or tubing
	Solenoid valve stuck open	If no voltage is on coil, replace solenoid valve
	Lost fill of non-fail-safe bellows	Replace control
	Termination setting too high	Correct setting
Noisy control	Control mounted in wrong position	Correct mounting position; if still noisy replace control
	Noisy clock motors	Replace control

OIL BURNER
SERVICE SUGGESTIONS

TROUBLE	CAUSE	REMEDY
Repeated safety shutdown	Inadequate combusion detector response	If no response, replace detector. If response is inadequate, move detector to respond to stable part of flame
	Low line voltage	New wiring, or contact power company
Short cycling of burner	Dirty filters	Clean
	Faulty operation of auxiliary controls	Reset, repair, or replace auxiliary controls
	Incorrect thermostat anticipation	Set to higher amperage
Relay will not pull in	No power. Open power circuit	Reset, repair, or replace
	Open thermostat circuit	Check thermostat wiring

ELECTRIC CONTROLS
FOR REFRIGERATION
AND AIR CONDITIONING

WORKBOOK

To power supply

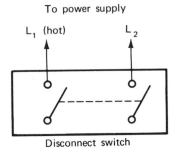

Disconnect switch

Compressor motor

Connect the power source through the disconnect switch to the compressor motor so that it will operate when the switch is closed.

To three phase
power supply

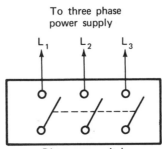

Disconnect switch

Compressor motor

Connect the power source through the disconnect switch to the compressor motor so that it will operate when the switch is closed.

To power supply

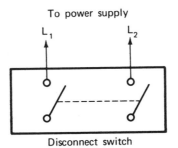

Disconnect switch

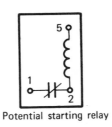

Potential starting relay

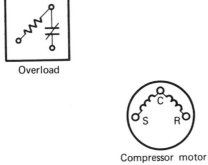

Overload

Compressor motor

Complete the diagram to include the starting relay and over-load.

To power supply

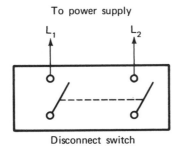

Disconnect switch

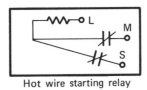

Hot wire starting relay

Compressor motor

Complete the wiring diagram including the hot wire relay.

To power supply

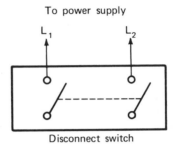

Disconnect switch

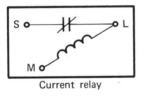

Current relay

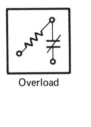

Overload

Compressor motor

Complete the diagram including all components.

To power supply

L₁   L₂

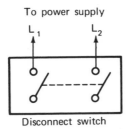

Disconnect switch

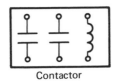

Contactor

Overload

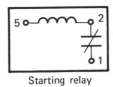

Starting relay

Cold control
(thermostat)

Starting capacitor

Compressor motor

Complete the diagram including all components in the line voltage circuit.

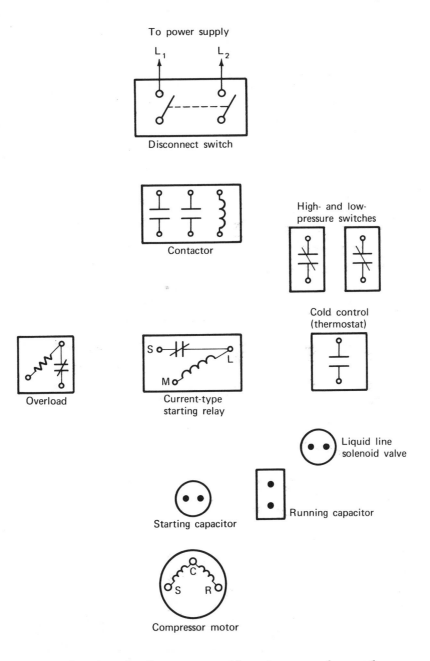

To power supply

L₁   L₂

Disconnect switch

Contactor

High- and low-
pressure switches

Cold control
(thermostat)

Overload

Current-type
starting relay

Liquid line
solenoid valve

Starting capacitor

Running capacitor

Compressor motor

*Complete the diagram to provide system pump-down and
high pressure safety.*

To three phase
power supply

L₁   L₂   L₃

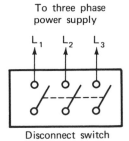

Disconnect switch

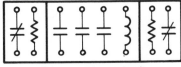

Starter with overloads

High- and low-
pressure switches

Cold control
(thermostat)

Liquid line
solenoid valve

P₁

P₂   P₃

Compressor motor

*Complete the diagram to provide system pump-down and
high pressure safety. Include all components.*

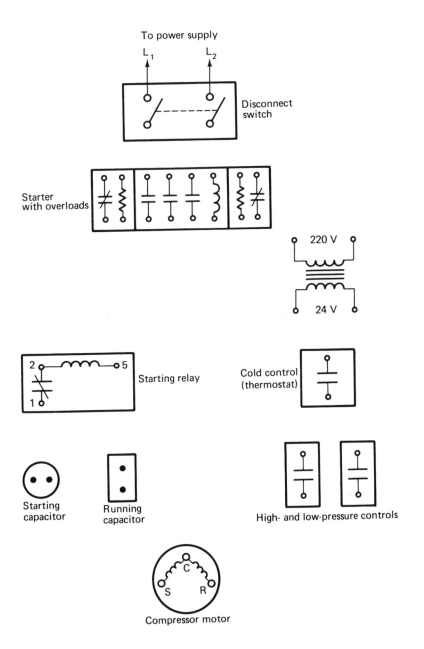

To power supply

L₁   L₂

Disconnect switch

Starter with overloads

220 V

24 V

Starting relay

Cold control (thermostat)

Starting capacitor

Running capacitor

High- and low-pressure controls

Compressor motor

Complete the diagram, placing all controls possible in the low voltage circuit.

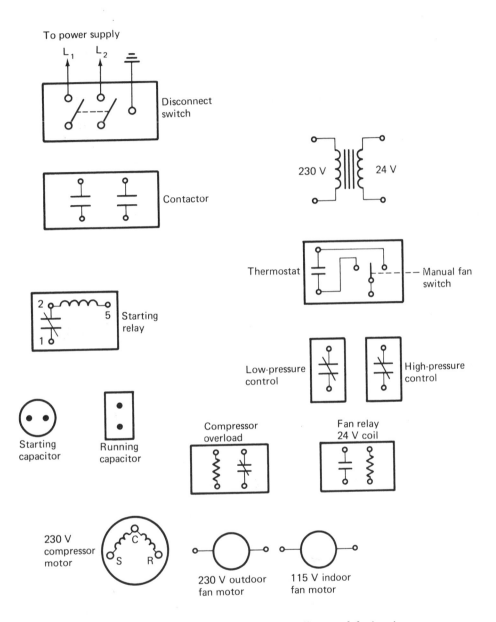

To power supply

L₁ L₂

Disconnect switch

Contactor

230 V   24 V

Thermostat — — Manual fan switch

2 5 Starting relay

Low-pressure control   High-pressure control

Starting capacitor   Running capacitor

Compressor overload   Fan relay 24 V coil

230 V compressor motor

230 V outdoor fan motor   115 V indoor fan motor

Complete this cooling diagram placing all control devices in the low voltage (24 V) circuit. Use a red pencil for voltage, a green pencil for the fan circuit, and yellow for cooling.

To power supply

Fan control

115 V indoor fan motor

Thermostat

Pilot safety device

Limit control

Main gas valve, 24 V

Complete the 24 V heating diagram.

To power supply

Fan control

115 V indoor fan motor

750 mV powerpile

mV thermostat   Limit control   mV gas valve

Complete the above 115 mV heating diagram.

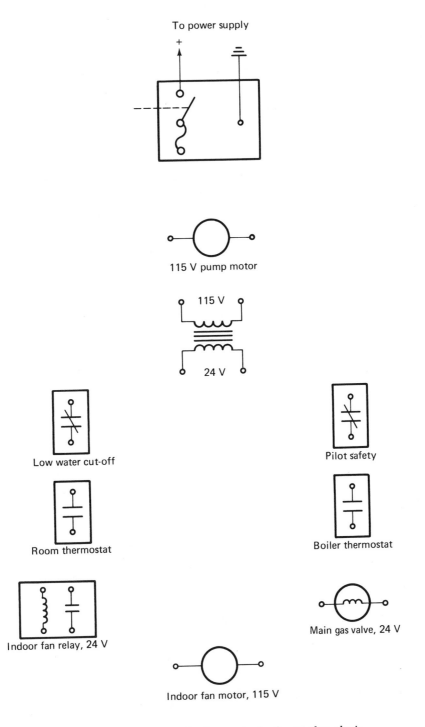

To power supply

115 V pump motor

115 V

24 V

Low water cut-off

Pilot safety

Room thermostat

Boiler thermostat

Indoor fan relay, 24 V

Main gas valve, 24 V

Indoor fan motor, 115 V

Complete the diagram of a hot water boiler so that the indoor thermostat controls the indoor fan and the boiler thermostat controls the gas valve through the safety devices.

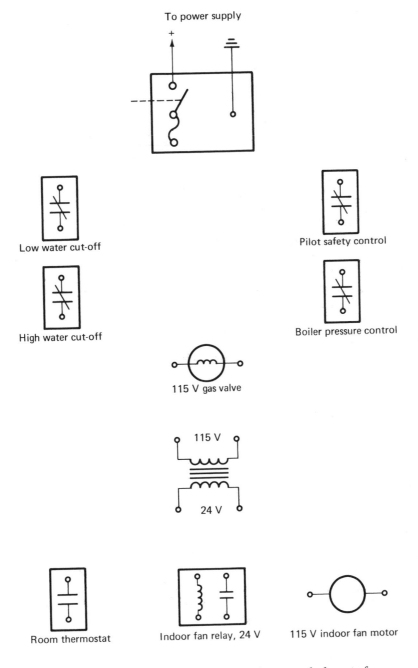

To power supply

Low water cut-off

High water cut-off

Pilot safety control

Boiler pressure control

115 V gas valve

115 V

24 V

Room thermostat

Indoor fan relay, 24 V

115 V indoor fan motor

Complete the diagram so that the boiler controls do not affect the indoor controls.

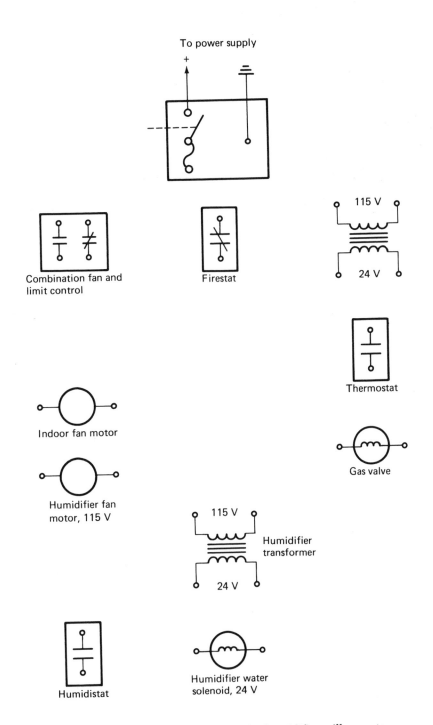

To power supply

Combination fan and
limit control

Firestat

115 V

24 V

Thermostat

Gas valve

Indoor fan motor

Humidifier fan
motor, 115 V

115 V

Humidifier
transformer

24 V

Humidistat

Humidifier water
solenoid, 24 V

*Complete the diagram so that the humidifier will operate
only when the indoor fan is operating.*

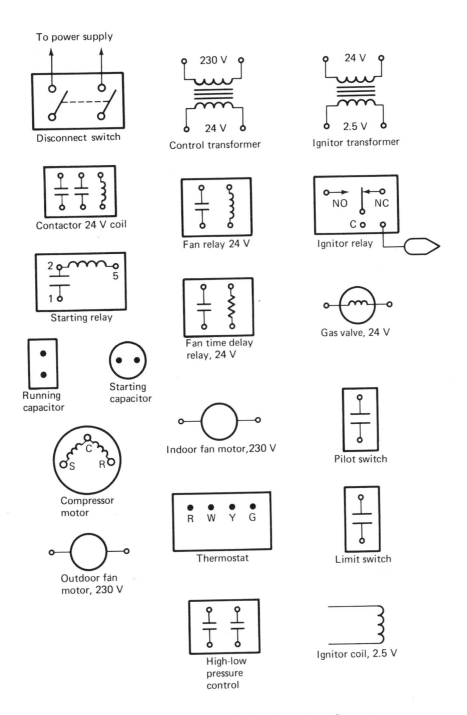

To power supply

Disconnect switch

230 V
24 V
Control transformer

24 V
2.5 V
Ignitor transformer

Contactor 24 V coil

Fan relay 24 V

NO   NC
C
Ignitor relay

2   5
1
Starting relay

Fan time delay
relay, 24 V

Gas valve, 24 V

Running
capacitor

Starting
capacitor

Indoor fan motor, 230 V

Pilot switch

C
S   R
Compressor
motor

R   W   Y   G
Thermostat

Limit switch

Outdoor fan
motor, 230 V

High-low
pressure
control

Ignitor coil, 2.5 V

*Complete the above diagram of a heating and cooling sys-
tem with an automatic pilot ignitor. Use map pencils col-
ored: R—voltage; W—heating; Y—cooling; G—fan.*

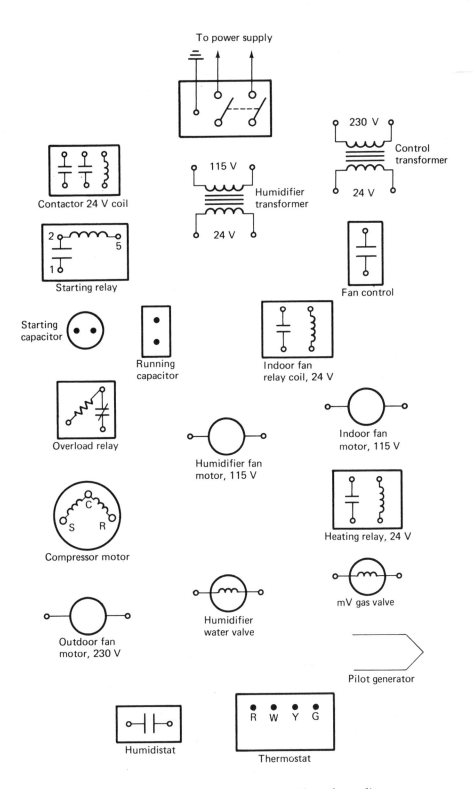

Complete the above diagram, following the color coding as set before. Do not *mix the voltages.*

To power supply

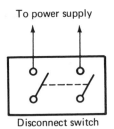

Disconnect switch

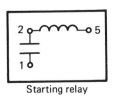

Starting relay

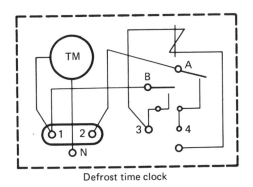

Defrost time clock

Compressor motor

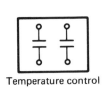

Temperature control

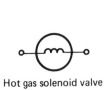

Hot gas solenoid valve

Fan motor

Complete the diagram of a simple hot gas defrost system.

To the
power supply

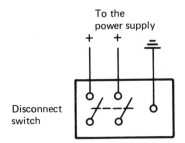

Disconnect
switch

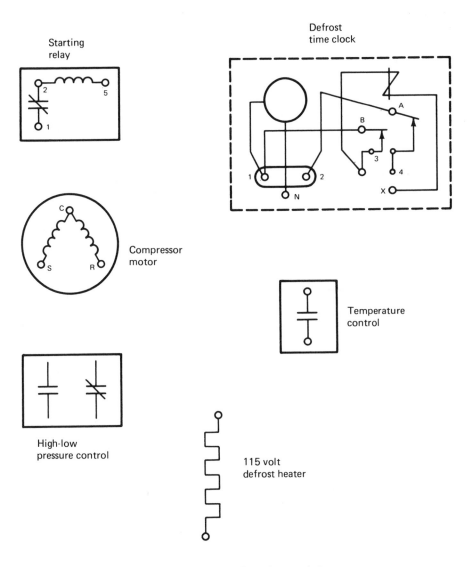

Starting
relay

Defrost
time clock

Compressor
motor

Temperature
control

High-low
pressure control

115 volt
defrost heater

Complete the above diagram of an electric defrost system.

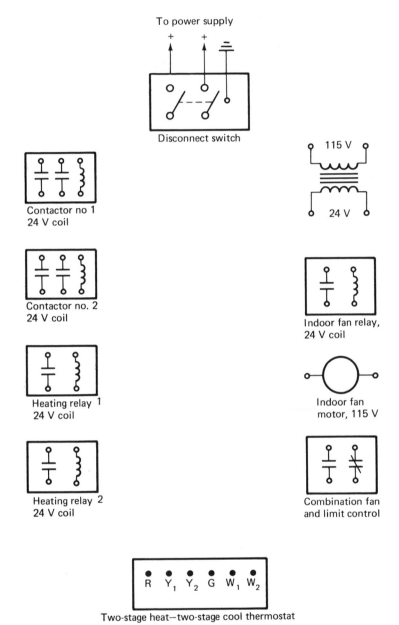

To power supply

Disconnect switch

Contactor no 1
24 V coil

Contactor no. 2
24 V coil

Heating relay 1
24 V coil

Heating relay 2
24 V coil

115 V

24 V

Indoor fan relay,
24 V coil

Indoor fan
motor, 115 V

Combination fan
and limit control

R Y₁ Y₂ G W₁ W₂

Two-stage heat—two-stage cool thermostat

Complete the diagram for two-stage heating and two-stage cooling. Use the color code.

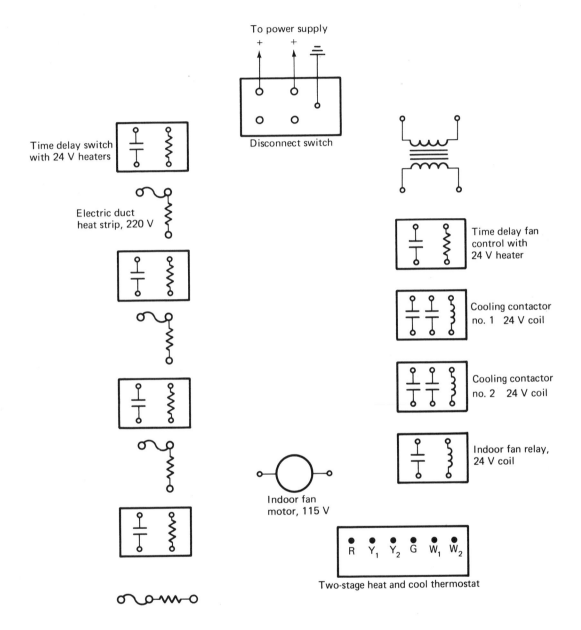

Time delay switch
with 24 V heaters

Electric duct
heat strip, 220 V

To power supply
+   +

Disconnect switch

Time delay fan
control with
24 V heater

Cooling contactor
no. 1   24 V coil

Cooling contactor
no. 2   24 V coil

Indoor fan relay,
24 V coil

Indoor fan
motor, 115 V

R   Y₁   Y₂   G   W₁   W₂

Two-stage heat and cool thermostat

Complete the diagram, wiring one-half the heat strips on each thermostat stage. Follow the color code.

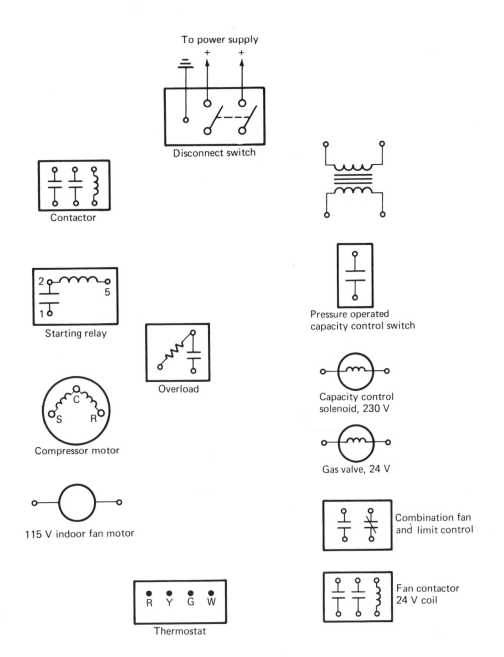

To power supply

Disconnect switch

Contactor

Starting relay

Overload

Compressor motor

115 V indoor fan motor

Thermostat

Pressure operated
capacity control switch

Capacity control
solenoid, 230 V

Gas valve, 24 V

Combination fan
and limit control

Fan contactor
24 V coil

*Complete the above diagram, which includes compressor
capacity control.*

INDEX